IN THE COUNTRY OF THE KAW

In the Country of the Kaw
A Personal Natural History of the American Plains

James H. Locklear

UNIVERSITY PRESS OF KANSAS

© 2024 by the University Press of Kansas
All rights reserved

Parts of Chapter 5 appeared previously in different form as "Shaggy Grass Country" in *NEBRASKAland* 75 (June 1997): 8–15. Reprinted with permission of the Nebraska Game and Parks Commission.

Published by the University Press of Kansas (Lawrence, Kansas 66045), which was organized by the Kansas Board of Regents and is operated and funded by Emporia State University, Fort Hays State University, Kansas State University, Pittsburg State University, the University of Kansas, and Wichita State University.

Library of Congress Cataloging-in-Publication Data
Names: Locklear, James H., 1953- author.
Title: In the country of the Kaw : a personal natural history of the American plains / James H. Locklear.
Description: Lawrence, Kansas: University Press of Kansas, 2024 | Includes bibliographical references and index.
Identifiers: LCCN 2023044342 (print) | LCCN 2023044343 (ebook)
ISBN 9780700636419 (cloth)
ISBN 9780700636426 (paperback)
ISBN 9780700636433 (ebook)
Subjects: LCSH: Natural history—Kansas—Kansas River Watershed. | Watershed ecology—Kansas—Kansas River Watershed. | Kansas River Watershed (Kan.)—Environmental conditions | Kansas River Watershed (Kan.)—Description and travel. | Locklear, James H., 1953- | Kansas River Valley (Kan.)—Biography. | BISAC: NATURE / Ecosystems & Habitats / Plains & Prairies | SCIENCE / Natural History
Classification: LCC QH105.K3 L63 2024 (print) | LCC QH105.K3 (ebook) | DDC 578.09781/3—dc23/eng/20231212
LC record available at https://lccn.loc.gov/2023044342.
LC ebook record available at https://lccn.loc.gov/2023044343.

British Library Cataloguing-in-Publication Data is available.

Printed in the United States of America

10 9 8 7 6 5 4 3 2 1

The paper used in this publication is acid free and meets the minimum requirements of the American National Standard for Permanence of Paper for Printed Library Materials Z39.48-1992.

For Marie

They had best understood that the soul of a landscape is a story, and the soul of a story is a personality.

G. K. Chesterton, *The Everlasting Man*

Contents

Acknowledgments xi

Prologue xv

Part I The Face of the Land 1

1 Headlands 3

2 Heartland 14

3 Scarplands 25

4 Tree Folk 37

5 Setting Sail on the Star-Grass Sea 50

6 A Kaw Florilegium 63

Part II The Membership 75

7 The Days of Manure River 77

8 O Elkader! 88

9 Bird Sketches 100

10 Surprised by Shorbs 113

11 The Waters of Mother Kaw 125

12 Beautiful Contrivances 138

13 The Hard Places 150

Part III Nature & Culture 161

14 Paradise Undone 163

15 Community Ecologies 176

16 The Showalter Lilac   188

17 Lithophilia   196

18 The Greatest Day in the History of Beeler   209

19 Rock Towns   218

20 Stickers   231

Epilogue   243

Appendix A: Plants Mentioned in the Text   247

Appendix B: Animals Mentioned in the Text   253

Selected Bibliography   259

Index   287

A color photograph gallery follows page 230.

Acknowledgments

Thanks to my employers over the past three decades—the Dyck Arboretum of the Plains in Hesston, Kansas, the Nebraska Statewide Arboretum based at the University of Nebraska-Lincoln, and Lauritzen Gardens in Omaha, Nebraska. Many of the observations and experiences recorded in this book happened while traveling around the plains on "business." I'm proud to have worked for each of these great institutions.

Thanks to the folks I met along the way who helped me to see hidden wonders or to better understand what I did see. I can't name all of you, but need to call out a few who made special contributions to particular chapters:

Heartland—Robert Blazing, retired archaeologist with the US Department of the Interior Bureau of Reclamation, for sharing personal research that enabled me to find the Guide Rock.

Scarplands—Donna Wilcox, daughter of Bertrand and Marian Schultz, for sharing information about the use of a glacial boulder to mark the graves of her parents in the Red Cloud (Nebraska) Cemetery.

Tree Folk—Jack Phillips, Robert Smith, and Bruce Hoffman, three fairly obsessed "oak people," for insight into the ecology and location of bur oak stands in western Nebraska and Kansas. Tom Maupin for touring me around the wonderful stand of bur oak on his family ranch on Paradise Creek in Russell County, Kansas.

Bird Sketches—The Manuscripts Collection of the Washington State Library for a photocopy of the Nathan Pattison diary.

Surprised by Shorbs—Joel Jorgensen, Nongame Bird Program Manager with the Nebraska Game and Parks Commission, for conversations about shorebird ecology in the Rainwater Basin. Dr. John McCarty and Dr. LaReesa Wolfenbarger, professors of biology at the University of Nebraska-Omaha, for sharing their experiences with shorebirds in the

Rainwater Basin and letting me join in their annual buff-breasted sandpiper survey.

The Waters of Mother Kaw—Craig Pruett, proprietor of Up a Creek Canoe and Kayak Rental in Lawrence, Kansas, for coining the phrase "Mother Kaw." Mark Eberle (retired, Fort Hays State University), Dr. Keith Gido (Kansas State University), Dr. Erika Martin (Emporia State University), and Joe Tomelleri (artist) for sharing expertise on aquatic ecology. Special thanks to Mark for his careful review of the manuscript for this chapter, and Joe for making me aware of the Fish Wizards Tour in the Flint Hills.

The Showalter Lilac—Ronald Dietzel, volunteer archivist par excellence with the Harvey County Historical Society in Newton, Kansas, for historical research into the story behind the Showalter lilac. Dawn McInnis and Jamie Rees of the Clendening History of Medicine Library at the University of Kansas Medical Center for historical research on the diseases of pioneer-era children.

Lithophilia—Richard Eckles and Tom Westfall, highly respected avocational archaeologists, for allowing me to view their personal artifact collections and for insightful conversations about lithic resources and transport in the Great Plains. Special thanks to Tom for bringing to my attention quotes from James Michener's *Centennial*. Dr. Steven Nash and Amy Gillaspie of the Denver Museum of Nature & Science for facilitating access to the Jones-Miller artifacts. Pete Felton Jr., for reminiscing about his life and work as a sculptor during a visit to his studio in Hays, Kansas, on October 12, 2021.

Stickers—Dawn Buehler (Friends of the Kaw), Bill and Jan Whitney (Prairie Plains Resource Institute), and Nathan and Laura Andrews (The Nature Conservancy's Fox Ranch Preserve) for sharing their stories and passion. Chris Pague, senior conservation ecologist with The Nature Conservancy, for facilitating access to the Fox Ranch Preserve and for insight into conservation issues in the Great Plains.

Epilogue—Doug Stevenson, senior surveyor of the Holdrege, Nebraska, office of Olsson for determining the exact height of the former power plant chimney at the site of Camp Atlanta.

Thanks to the talented photographers who made their work available for this book: Scott Bean, Michael Forsberg, Eric Fowler, Bruce L. Hogle,

William C. Johnson, Joel G. Jorgensen, Harland J. Schuster, Dave Showalter, and Craig Thompson.

Thanks to the University Press of Kansas for agreeing to this unusual project and to reviewers Rex Buchanan and Kelly Kindscher, who made the book better. Special thanks to Joyce Harrison, editor in chief, for guiding me through the whole publishing process.

Thanks to my brother, Jeff, my travel companion on many trips across I-70 and my guide to the mysteries of the West Bottoms. And here's to our folks, Ed and Mary Ann Locklear!

Thanks to my kids, Kendra, Karen, and Greg, now adults with beautiful families of your own. So many fond memories of hikes, picnics, camping, and day trips to see the sometimes-quirky wonders of the country of the Kaw. Thanks for humoring me back then and for encouraging me now.

Finally, thanks to my wife, Lynn, a deeply rooted woman in every way—a tree planted by streams of water. I love you and admire you and I'm so thankful for the life we've had together.

Prologue

> I see that the life of this place is always emerging beyond expectation or prediction or typicality, that it is unique, given to the world minute by minute, only once, never to be repeated. And then is when I see that this life is a miracle, absolutely worth having, absolutely worth saving.
> *Wendell Berry,* Life Is a Miracle: An Essay against Modern Superstition

I have two bird fantasies, both entangled with the Kaw, one unattainable in this life, the other on my bucket list.

The first involves the Carolina parakeet, a parrot-like bird of green, yellow, and red plumage. About the size of a pigeon, Carolina parakeets once ranged in spectacular, boisterous flocks across the eastern United States, especially favoring wooded river valleys. Lewis and Clark encountered them while in camp at the confluence of the Kaw River with the Missouri River on June 26, 1804, with William Clark journaling, "I observed great numbers of Parrot queets this evening." His report of Carolina parakeets in what would become the industrial heart of Kansas City was the first sighting of this species west of the Mississippi.

The story of the Carolina parakeet is one of the great sorrows of American environmental history. As more of the eastern United States was settled and the land converted to agriculture, these birds acquired a taste for grain crops and orchard fruit. Because of their staggering numbers, Carolina parakeets could inflict heavy losses. Eradication efforts were launched and, coupled with habitat loss and market hunting for their colorful feathers, the Carolina parakeet was wiped from the face of the Earth by the start of the 1920s.

It is easy for someone today to condemn the actions of folks living in an agrarian society over a century ago. But I don't have to labor, day in, day out, year after year, to raise crops to feed my family and pay my bills.

And I imagine that given their enormous numbers no one expected these birds would go extinct, even if they wished they would just disappear locally.

Still, Clark's brief journal entry haunts me. I can't help but dream about the sight—a green vortex of Carolina parakeets churning over the mouth of the Kaw!

The prophet Isaiah speaks of a day when this beat-up Earth will be made new. I wonder if Carolina parakeets will be back in the mix. My guess is, it would make God smile to bring them back.

My other bird fantasy springs from the vicinity of Cope, Colorado. Once, when doing an internet search about birding in eastern Colorado, I came across a tantalizing post: "At 11 am today (9/20/2010), a kettle of about 70 Swainson's hawks soaring over Cope, CO, Washington County, then quickly drifted off to the east." *A kettle of Swainson's*. . . . I felt an ache, something between longing and jealousy, rising in my heart.

The village of Cope, population 169, is located on the Arikaree River, the westernmost tributary of the Kaw, about 450 miles out from Kansas City as the crow flies. If not for the even smaller Anton, another twenty miles beyond, Cope would be the westernmost community in the entire watershed of the Kaw. To the north of Cope are rolling sandhills covered with sandsage prairie; to the south is High Plains tableland, some of it in crops, some of it still in shortgrass prairie. Prime habitat for Swainson's hawks.

Swainson's are true grasslanders, taking over the skies from midwestern red-tailed hawks as you move west and the trees become confined to waterways. Swainson's have the longest annual migration of any American raptor, an astounding ten-thousand-mile round trip between the Great Plains and the grasslands of Argentina. Arriving on the High Plains in April, they start the trip back to South America in September. That's where the kettle comes in.

This picturesque term refers to a spiraling flock of hawks, rising high in the sky, circling en masse while waiting to catch thermals to carry them along on their migration. Not all species of hawks do this, and apparently none as spectacularly as Swainson's. There are reports of Swainson's hawks wheeling above the late summer plains in kettles of hundreds, even thousands of birds.

I've seen plenty of Swainson's hawks over the years, but I've never seen them gathered in a kettle. I imagine it is like a colossal thunderhead that suddenly gets up over the plains: something you can't really plan on seeing; you just have to be in the right place at the right time.

Note to family: If I turn up missing from the old folks' home some September day, you might try looking for me out on the Arikaree.

Between these two avian dreamscapes lies the subject of this book. You can never know all there is to know about a place, but after a half-century of rambling and rooting around in the realm of the Kaw River, I feel compelled to write down what I have seen. I will tell you now, the country of the Kaw is like no other place on Earth, and like every other place on Earth.

Astonishments abound!

The Missouri River is the third longest on the planet and has been called "the master stream" of the American Great Plains. Its largest tributaries are the Milk, Yellowstone, and Platte Rivers, all mountain born, and the Kaw, which alone rises on the High Plains. Entering the Missouri just as it makes its monumental swing from south to east, the Kaw is the last stream off the plains as the Big Muddy sets its sights on St. Louis.

The Kaw River and the Kansas River are one and the same. The latter is the more proper name, the one found in most government and academic publications. The former is what a geographer would call a vernacular name—the one the locals use. Like your nickname for an old friend, there is more emotion and affection behind it.

The name of the river derives from what Euro-Americans understood to be the name of the Indian peoples who were the dominant tribe in the eastern reaches of the river's watershed. French Canadian fur traders were first to take a crack at it, with *Kanza* appearing among a list of tribal names on a map prepared in 1674 by voyageur Louis Jolliet. The first application of the name to the river is found on the 1718 map of cartographer Guillaume Delise, who rendered it *Grande Riviere des Cansez*. American explorers later took up these and similar-sounding French names and anglicized them to Kanses, Kanzas, or Kanzus.

In researching *PrairyErth*, his sweeping epic on the Flint Hills, William Least Heat-Moon came across 140 variations of the word *Kansas*. He lists them all, in four columns of type. One is *Kaw*, which is the preferred tribal name of the modern descendants of the Kansa people. It is also the favored name of everyday Kansans for what is rendered *Kansas River* on their official state highway map. The earliest documented use I can find is a qualified "Kaw River or Kansas R." on a map prepared in 1857 under the supervision of Lt. G. K. Warren of the US Topographical Engineers.

Preference for "Kaw River" is particularly pronounced among Kansas Citians, but also circulates in Lawrence and Topeka. Innumerable businesses in these river towns have "Kaw" or "Kaw Valley" in their names, from banks to bars to dental clinics to sand and gravel suppliers. And there's Kaw Point at Kawsmouth, where William Clark spotted all those Carolina parakeets. And Friends of the Kaw, a nonprofit organization dedicated to preserving and protecting the river.

I'm going with "The Kaw," since I grew up in Kansas City and it rolls off the tongue better than "The Kansas." Whatever you call it, the actual stretch of river so-named has a fairly modest run, 173 water miles from Junction City, Kansas, to confluence with the Missouri. But there is much more to its domain. The Kaw is birthed on the east edge of Junction City from the comingling of waters from the Republican and Smoky Hill—streams that share headwaters on the distant High Plains of eastern Colorado. And several other major tributaries enter the Kaw on its way east from this juncture, notably the Big Blue coming down from Nebraska, and the Vermillion, Wakarusa, and Delaware, native-born Kansas streams. All told, sixty-one thousand square miles of the Earth's surface lies under the sway of the Kaw River.

I have not always been a person who daydreams about birds. But the bent of my soul toward such imaginings was set early in life. Ed and Mary Ann Locklear moved their sons Jeff and Jim from South Bend, Indiana, to Kansas City, Missouri, in March 1962. Jeff was eight, I was nine. Our brand-new ranch-style house was located on what was then the far south edge of KC, in a brand-new neighborhood full of other young families, most of these populated by boys. Just a couple of blocks away was not-yet-developed countryside laced with woodlands, creeks, and pastures. It was like we had arrived on the frontier.

I was fortunate to have had the kind of "free-range" childhood described by Richard Louv (also a Kansas City kid) in *Last Child in the Woods*, and my buddies and I spent lots of time roaming these edge-of-town wildlands without any adult supervision. It wasn't that our parents were negligent; in those days there was just a shared understanding—maybe even a code among the moms—that the best thing for boys was to get them out of the house and let them be boys. We were too young for sports and too clueless about girls to be distracted by them, and organized youth activities were limited, so we were left to make our own fun in the Great Outdoors.

If we had most of the day ahead of us, as on a weekend or during summer, we would head to the Blue, sometimes called the Big Blue River. To be clear, this is not the Big Blue River of the Kaw watershed, which gathers its waters in south-central Nebraska and joins the Kaw east of Manhattan, Kansas. The Blue of my boyhood is a small stream that arises in the rocky, hilly country on the south side of the KC metro and angles north-northeast about forty miles to its confluence with the Missouri River, nine miles downstream from Kaw Point. Its headwaters are in Johnson County, Kansas, as are those of Cedar Creek, Mill Creek, and Turkey Creek, the last three streams to enter the Kaw before it reaches the Missouri.

Our expeditions to the Blue were on foot or by bike, and the section best known to us was from the Red Bridge crossing near Minor Park south to Martha Truman Road. While the lower part of the river is little more than an engineered trench as it makes its way through the postindustrial East Bottoms of the Missouri, our stretch coursed across a rocky streambed bordered by sycamores below limestone bluffs cloaked with oak-hickory forest. Thankfully, that run is still in good condition today; the Jackson County Parks and Recreation Department holds most of the adjacent land and maintains a system of hiking trails along the river.

We rarely had thought-out plans for these excursions; we just headed to the river to see what was up. It became harder to organize a group foray as we entered our teen years, but I still found myself drawn to the river and its woods, even if it meant going alone. I don't recall a specific aim of these solo trips either, but the solitude enabled me to notice more of the detail around me. Peterson field guides and other nature books helped me put names to the plants and birds and other creatures I was seeing. The more I learned, the more enchanted I became. I believe it was this boyhood taste of magic that in later life made me susceptible to bird fantasies.

I blame it on the Blue.

Ed Locklear was not an outdoorsy man. My dad was an executive with a life insurance company, who left our house early in the morning in a tailored Jack Henry suit to work in a fancy office on the twentieth-something floor of the Power and Light Building in downtown Kansas City. He never camped or went for a hike in the woods, and he was convinced that the cigarette butts he tossed from the car window would be used by birds to build their nests.

But he was crazy about fishing. Locklear family vacations were always glorified fishing trips to places Dad discovered in the pages of *Field & Stream* and *Outdoor Life*. For many years we would drive north to some lake in Minnesota to pursue walleye, smallmouth bass, and northern pike. Then one year, probably after we experienced the misery of blackfly hatch in the North Woods, Dad decided we would head west, to fish for trout in the Colorado Rockies. We never went to Minnesota again.

The Rocky Mountains are exhilarating, especially the first time you experience them. But I also found something strangely stirring about the roofless landscape we traversed between the oakiness of Kansas City and the pineyness of the mountains. The more times we went to Colorado, the more I would look forward to the next drive that got us there. Little did I know, but each trip along Interstate 70 was a five-hundred-mile-long sampling of the country of the Kaw.

I graduated from Center High School in 1972 and that fall started college at Central Missouri State University in Warrensburg, Missouri, about an hour east of Kansas City. During my sophomore year, Dad took a job in Denver—a great career move that also put him within sight of his now-beloved mountains and their trout streams. My brother Jeff was also in school at Central Missouri, so going home for the holidays or summer break now meant even more drives across I-70.

The best thing to happen to me in college was meeting Lynn Marie North, who became my wife in 1977. Over the ensuing decades we lived in Kansas City; Carbondale, Illinois; Lincoln, Nebraska; Hesston, Kansas;

and then back in Lincoln. Kendra, Karen, and Greg were born to us in the 1980s. With us residing on the east side of the Great Plains and my folks on the other side in Denver and, later, Montrose, Colorado, there were many more trips through the Kaw watershed, first just Lynn and me, later with our kids in tow. With every crossing another subtle impression of the landscape was laid down, like the imperceptible deposits that eventually yielded the Cottonwood, the Greenhorn, and other great limestones of the country of the Kaw.

The next best thing to happen to me in college was studying botany under David Castaner. Dr. Castaner was a gifted teacher with a quirky sense of humor and a barely contained childlike delight in the world of plants. He recognized my interest and let me work in the college herbarium and do a special project on lichens. All this stoked the fire kindled during my boyhood wanderings and precipitated a career path that led first to a garden center, then to university research greenhouses, and finally into the rarefied world of the arboretum and botanical garden.

Fortunate am I to have worked for three wonderful institutions dedicated to connecting people with plant life: the Dyck Arboretum of the Plains in Hesston, Kansas, the Nebraska Statewide Arboretum, and Lauritzen Gardens in Omaha. The nature of my work has sometimes been horticultural, as in collecting wildflower seed and growing plants for introduction into the green industry. And it has sometimes been botanical, as in searching for imperiled plants in the wild to understand their conservation needs. Thirty-some years of this have afforded me opportunities for deep soundings throughout the country of the Kaw, experiences mediated by plants but often veering off into other realms, where I could tug at other threads in the fabric of the cosmos.

I have in my possession a letter dated December 27, 1968, from Richard Brenneman, administrative officer in the Pollution Control Department of the City of Kansas City. Mr. Brenneman was writing in response to a letter penned by my fourteen-year-old self to the Missouri Water Pollution Board in Jefferson City about a small trash dump the guys and I discovered along the west bank of the Blue River.

My report had been referred to the Division of Waste Surveillance, which sent out an investigator who confirmed the existence of the dump.

Inquiry was made, and a resident of the area admitted to dumping trash at the site and said he would cease immediately. I also received a nice note from Jack K. Smith, executive secretary of the Missouri Water Pollution Board, commending me and my friends for caring so much about the Blue and encouraging us to "consider a field related to water pollution control."

I have no idea what I actually wrote in my letter, much less where I got the nerve to write it. I had no scientific facts or statistics to back up my appeal. I must have just aimed for the heart.

Now, more than fifty years later, I find myself presuming to speak on behalf of another landscape. The country of the Kaw is far more sweeping and its story far more complicated. But the same mighty sentiments are at work: awe, affection, sorrow, and hope; hope that those in a position to care for the land and its communities, natural and human, will see its goodness and strive for its well-being.

I'm still aiming for the heart.

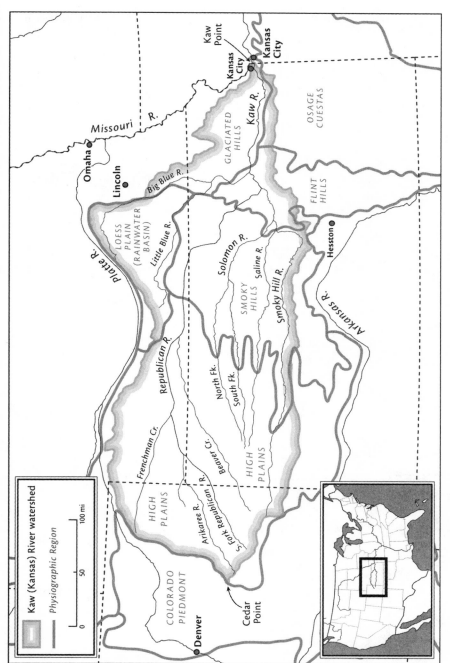

Watershed of the Kaw River showing physiographic regions and major streams.

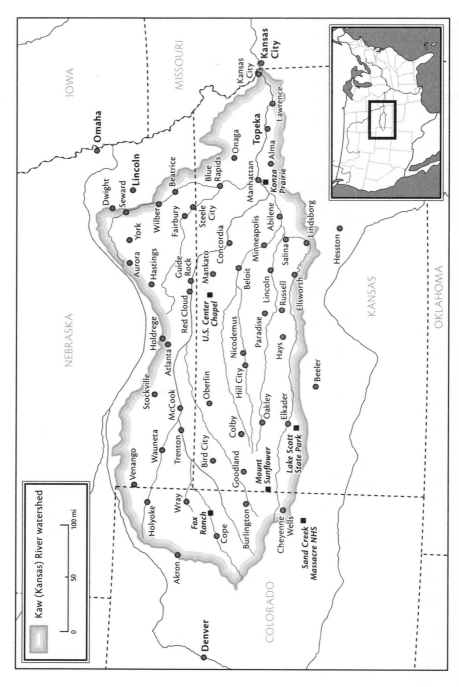

Selected place-names within the Kaw River watershed.

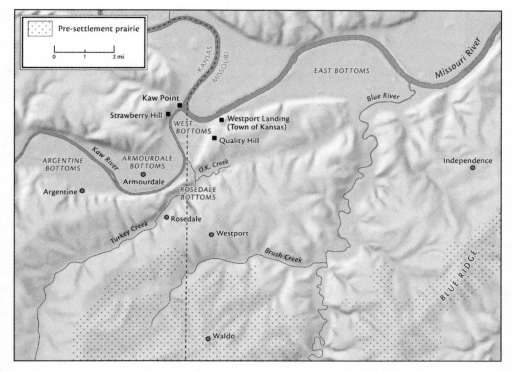

River bottoms, presettlement prairie, historical channels of Turkey and O.K. Creeks, and selected place-names associated with the Kaw River in the Kansas City metropolitan area. Modified from a map in James Shortridge, *Kansas City and How It Grew, 1822–2011* (Lawrence: University Press of Kansas, 2012), p. 5, with additional data from a map in Walter A. Schroeder, "The Presettlement Prairie in the Kansas City Region," *Missouri Prairie Journal* 7 (December 1985): 4.

PART I

The Face of the Land

> In the same way that one discerns a certain physiognomy
> in individual organic beings . . . so too is there a
> physiognomy of Nature that applies, without exception,
> to each section of the Earth.
> *Alexander von Humboldt,* Views of Nature

Alexander von Humboldt was what was once called a *polymath*—an intensely curious person whose scholarly knowledge spanned an amazing breadth of subjects and disciplines. But he was a polymath with a poet's heart. Having traveled on four continents, it dawned on the great German explorer-naturalist that landscapes can have a distinct visual personality—a face. He proposed the daring idea of a physiognomy of nature in his works *Views of Nature* and *Essay on the Geography of Plants*, both published in German in 1807. Like the impression captured by a painter, physiognomy reflects the unique, localized imprint of geology, land forms, and climate, which in turn influences the shape and composition of plant life, which is the most important visual signature of the environment—the face of the land.

ONE

Headlands

> The River Kanses takes its Source on these Plains.
> *Zebulon Pike, notation on* Chart of the
> Internal Part of Louisiana

Nevin Fenneman got excited about flatness. Writing in 1931 of the High Plains of western Kansas and eastern Colorado, the renowned geographer marveled at "the phenomenal flatness" of these "almost perfect plains," proclaiming the region "as flat as any land surface in nature." Such language is not likely to make it into a modern tourism campaign, but it speaks to the singularity of the High Plains landscape.

Flat it is. On average, there is an almost imperceptible elevation gain of ten to fifteen feet per mile from east to west across the High Plains; and in some places the surface is all but featureless, "virtually unscored by erosion." As defined by Fenneman, the High Plains stretches from southern South Dakota and southeastern Wyoming through the Texas Panhandle and eastern New Mexico.

It is on these flatlands that the journey begins for the big river that comes rolling into Kansas City, looking for the Missouri. The most far-flung tributaries of the Kaw arise on the High Plains of eastern Colorado, southwestern Nebraska, and northwestern Kansas and flow into either the Republican or Smoky Hill River, which eventually unite in the Flint Hills to form the Kaw proper. The headwater streams of the Republican are the Frenchman, North Fork, Arikaree, South Fork, Beaver, Sappa, and Prairie Dog. Those of the Smoky Hill are the Solomon, Saline, Hackberry,

North and South Forks, and Ladder. The High Plains dawning of the Kaw sets it apart from the other major tributaries of the Missouri River, all of which arise in the Rocky Mountains.

The astonishing flatness of the High Plains owes its origin to an extraordinary source—mountain rubble. Some seventy million years ago, western North America experienced a period of mountain building called the Laramide Orogeny that created a belt of mountains from Alaska to Mexico. Giant streams flowing eastward out of the Rocky Mountains were full of sand, gravel, and other rock debris, so much so that their sediments overflowed the meandering stream valleys and spread out over the uplands. By about ten million years ago, the deposits of these ancient streams had coalesced into one immense sheet of sediment that stretched in a gently sloping plain from the mountain front almost across what is today Kansas and Nebraska. This great wedge of transported mountain bits and pieces is the Ogallala Formation.

While "high plains" is used in a broad sense for the western part of the Great Plains, Fenneman and other geographers use the term "High Plains" more narrowly to refer to the region underlain by the Ogallala Formation and, north of the North Platte River, the older Arikaree Formation. The region is called "high" because stream action has gnawed into the eastern and western reaches of these formations, resulting in an escarpment-rimmed tableland that stands at a higher elevation than the surrounding erosion-cut landscape. This phenomenon is on dramatic display in the southern High Plains, where the cliff-and-canyon-bordered Llano Estacado—the palisaded plains—rises as an immense plateau above the surrounding terrain.

You can also see it at a place called Cedar Point, north of the town of Limon, Colorado. Cedar Point is a triangle-shaped westward-pointing promontory of tableland that stands several hundred feet above the plains below. The country between Cedar Point and the Rocky Mountain front is called the Colorado Piedmont. Here, the High Plains surface has been stripped away by streams flowing northward into the South Platte River and southward into the Arkansas. Cedar Point itself is the headwaters of the Arikaree River, the westernmost tributary of the Kaw and, at an elevation of six thousand feet, the highest point within the Kaw watershed, a mile closer to the stars than Kaw Point.

A number of windfarms are situated along the western edge of the

High Plains where Ogallala escarpments stand above the lower Colorado Piedmont, including the vast Cedar Point windfarm. This forest of 139 electricity-generating turbines is situated on twenty thousand acres of leased private land to the east of the brink of Cedar Point, and it looks for all the world like an advancing army of very agitated robots, especially at dusk when the red aircraft warning lights atop each turbine start flashing in unison.

The bulk of the Ogallala Formation consists of thick, poorly sorted sediments of sand and gravel derived from granite, sedimentary, and volcanic rocks. The upper zone of these sediments is often bound by a calcium carbonate cement called caliche or calcrete that forms by the evaporation of groundwater in this dry environment. These cemented beds have a concrete-like appearance and feel; where exposed by erosion, they form strong ledges called mortar beds. Outcrops of Ogallala mortar beds form the rimrock of escarpments and canyons and the caprock of buttes and hills in the High Plains.

At greater depths the coarse-textured Ogallala sediments are unconsolidated and relatively porous and permeable and are capable of holding tremendous volumes of groundwater. Here resides the Ogallala Aquifer, which makes up the bulk of the High Plains Aquifer system, the largest body of groundwater in the United States and a crucial resource for the modern economy of the Great Plains. The water held by the Ogallala Formation is fossil water, gathered from the melting of ancient mountain snow fields and glaciers.

Prior to Euro-American settlement, exposures of water-bearing Ogallala sediments in gullies, ravines, and canyons supported innumerable springs and seeps that were the primary fount of streams that head on the semiarid High Plains, where rainfall alone is usually too scant to sustain streamflow. Tragically, most of these have since dried up due to lowering of the water table by irrigation, and with their disappearance streamflow has in places been greatly diminished or vanished altogether.

Canyons are topographic anomalies in an otherwise dead-level landscape, providing islands and patches of resources not available on the open plains, plus protection from weather that at certain times of the year can kill you. A prime example is the area of broken country bordering what is today known as Ladder Creek, a tributary of the Smoky Hill River, in Scott County, Kansas. In addition to being an oasis of springs, firewood, and wild fruit, the canyons offer extraordinary concealment, being all but invisible from out on the surrounding plains, undetectable until you're standing right at the rim.

There is a great amount of natural and cultural history concentrated in this rugged country, which is today mostly encompassed by Lake Scott State Park. The area is home to the Scott riffle beetle, a very rare aquatic insect whose entire world is limited to a stretch of cool, highly oxygenated water in a spring-fed canyon-sheltered brook. Kansas antelope sage, a plant known from only a handful of places in western Kansas, occurs here as well, inhabiting rimrock along canyon edges.

It is also the setting of El Cuartelejo (also rendered El Quartelejo), a long-studied archaeological site that consists of the ruins of a seven-room, stone and adobe pueblo, and other artifacts. Known to archaeologists as the Scott County Pueblo, the site is important as evidence of contact and cultural exchange between the Puebloan peoples of the Sangre de Cristo Mountains of New Mexico and indigenous peoples of the plains. Historical knowledge of the area stretches back to the 1600s, when Spanish accounts record Indians from the Taos, and later Picuris, pueblos fleeing to El Cuartelejo to escape the rule of their colonial overlords.

Just south of the state park is the site of the Battle of Punished Woman's Fork, a branch of Ladder Creek where, on September 27, 1878, a band of Northern Cheyenne people under the leadership of Chief Dull Knife fought with US Army troops pursuing the Indians because of their unauthorized exodus from Oklahoma reservation lands. The canyons and bluffs provided an initial strategic advantage to the outmanned warriors as well as a place to shelter their women and children, and later gave cover as they slipped away under the cloak of darkness. The site is considered one of the few largely "pristine" Indian wars encounter sites in the nation.

That the Northern Cheyenne, whose homeland in today's Montana was five hundred miles to the northwest, and the Puebloan peoples of New Mexico, three hundred miles to the southwest, would head to this hidden cleft in the horizon at a time of trouble reveals remarkably intimate knowledge of the High Plains landscape by indigenous peoples, as well as deep esteem for these canyonlands.

I have always liked the feel of Wray, Colorado. While many towns and villages in the Great Plains are situated on the wide-open flats, Wray is tucked into the valley of the North Fork of the Republican, hemmed in on the north by sandhills and on the south by Ogallala-capped buttes and bluffs. Delightfully, a 1907 photo of one of the buttes overlooking Wray labels it Flirtation Rock. The town of Wauneta, Nebraska, is similarly situated in the bluff-bound valley of Frenchman Creek. A correspondent with the Federal Writers' Project of the Works Progress Administration wrote in 1939 that Wauneta "has the appearance of a mountain town."

A playa is an unpretentious landform befitting an understated landscape. Scattered across the surface of the High Plains, playas are round, shallow depressions lined with soils that have a higher clay content than surrounding soils. The origin of playas is thought to be some combination of subsidence, wind erosion, irregular soil formation, animal activities, or other factors, with the end result that the basin temporarily holds rainfall and snow melt, making playas ephemeral wetlands in a semiarid environment.

The High Plains has the greatest concentration of playas in the world, with the largest number occurring on the Llano Estacado of northwestern Texas and eastern New Mexico. Less well-known are the playas that occur on the High Plains of western Kansas and eastern Colorado, particularly about the headwaters of the Republican and Smoky Hill Rivers. Based on recent surveys in Colorado and Kansas, it is safe to say there are thousands of playas in the country of the Kaw. Most of these are modest in size, covering less than five acres, although a few are considerably larger.

Humble in appearance, playas collectively support biodiversity and groundwater recharge on the plains. Yet many playas have been filled so the ground within the basin can be farmed or have been pitted to concentrate water for irrigation pumping. Both practices alter the unique natural hydrology of playas and diminish the ecological services they provide. The conservation and restoration of Great Plains playas is a primary focus of the Playa Lakes Joint Venture, a partnership of federal and state wildlife agencies, conservation groups, private industry, and landowners.

The highest point in the state of Kansas is Mount Sunflower, elevation 4,039 feet, located in Wallace County, just a stone's throw from the Kansas-Colorado state line. Situated on the divide between two tributaries of the Smoky Hill River, Mount Sunflower is little more than a gentle rise in the horizontal High Plains landscape. But it is the Everest of adventurers with a taste for the ironic, who wisecrack about needing oxygen and technical climbing gear to reach the summit.

The view from Cedar Point—almost two thousand feet higher—is no joke. On a clear day you can easily see Pikes Peak, seventy-some miles away, and its steep cliffs produce updrafts strong enough to float ravens and golden eagles. The sight becomes dizzying when you consider that the High Plains tableland on which you're standing once extended all the way to the mountains, as did the ancestral Arikaree River. What happened to the land between? Geologists believe the mountain headlands of the ancestral Arikaree River, and thus the Kaw, were captured by tributaries of other streams, first the South Platte and then the Arkansas, leaving the Arikaree to gather its waters from what was left of its High Plains domain.

The hardest evidence for this theory has been found in the northwest corner of Kansas, where chunks of Pikes Peak granite occur in the canyonlands associated with the Arikaree River. Most of these stones are one to six inches in diameter, and it would require a much larger and more vigorous stream than the Arikaree of today to transport them nearly two hundred miles out from the Rockies. Even more remarkable, mixed in with these gravels are boulders eighteen to thirty inches in diameter that no stream or flood of water could have carried across the plains. The best guess is that these rocks were embedded in glacial ice and floated down from the mountains back when the Kaw, through the agency of the Arikaree, arose in the alpine.

I'm not sure where I first saw one, but you occasionally run into a handmade, tongue-in-cheek wind gauge in a High Plains small town park that consists of a heavy chain bolted to a post. The strength of the wind is indicated by the angle at which it holds the chain out from the post: 0°, which indicates no wind ("Gauge broken; notify meteorologist") to 30° ("Fresh breeze") to 90° ("Welcome to western Kansas").

High winds are synonymous with the High Plains and can pack a lot of dust and sand. These winds, orographic or terrain induced, accelerate by flowing downslope from the eastern front of the Rocky Mountains and then keep up a head of steam as the air rushes east across the flat, open terrain. Using the High Plains town of Burlington, Colorado, as an example, hourly wind speed averages 10.9 miles per hour from February into early June and peaks at 12.1 mph in April, the windiest month of the year. Sustained, non-thunderstorm gales of 30 to 40 mph or higher are not uncommon during the spring season.

But the winds of the past—paleowinds—were even stronger, and over time they transported enormous volumes of dust and sand through the air and across the land. In parts of the country of the Kaw, deep deposits of silt and sand blanket the High Plains surface and produce their own distinctive landforms. These are aeolian landscapes, created and shaped by the wind. The term derives from the Greek god Aeolus, keeper of the wind, said to reside on an island in the Mediterranean. Old Aeolus may take holidays in the Greek isles, but he spends most of the work week out on the High Plains.

Loess is strange stuff. Pronounced *luss*, it is composed of windblown silt particles, essentially tiny rock fragments, each smaller than a very fine grain of sand, cemented with calcium carbonate into a sedimentary formation. A chunk of loess is easily crumbled by hand, but a deposit of loess has the capacity to hold steep faces and vertical walls because, unlike the roundedness of sand grains, the angularity of the closely packed silt particles creates enough cohesion to resist erosion.

There are two major areas of loess deposition in the country of the Kaw. The eastern one occurs in the watersheds of the Big Blue and Little Blue Rivers in south-central Nebraska and adjacent Kansas. The landscape is a level to gently rolling plain of little topographic relief and very little scenic appeal unless you like the look of corn and soybean fields. It would be easy to dismiss this as the least interesting part of the country of the Kaw if not for the Rainwater Basin in Nebraska, where thousands of clay-bottomed depressions catch and hold precipitation, creating ephemeral wetlands of hemispheric importance to migrating shorebirds. More on that in a later chapter.

The second major area occurs to the southwest in the watershed of the Republican River. Here, deep deposits of loess mantle the High Plains surface, and the landscape has a lot more visual drama, consisting of intricately dissected tablelands and steep-sided, flat-bottomed draws that in some places expand into loess canyons. On August 5, 1873, a party of 1,500 Lakota warriors took advantage of this terrain when they surprised

350 Pawnee on a communal bison hunt near today's Trenton, Nebraska, slaying 69 men, women, and children in what would come to be known as Massacre Canyon.

Nebraska conservation scientists recognize a Loess Canyons Biologically Unique Landscape within this area, 530 square miles of steep loess hills and canyons south of the Platte River in Lincoln, Dawson, and Frontier Counties. Uplands here are cloaked with extensive stands of loess mixed-grass prairie, a distinctive grassland type of conservation concern due to the spread of eastern red cedar trees out of the canyons and into the rangeland. This picturesque country is drained by south-flowing tributaries of Medicine Creek, most of which run through canyons, and by Deer Creek, which has carved a highly scenic landscape along the eastern edge of the region.

Even more rugged canyons occur in southwestern Nebraska and adjacent parts of Kansas and Colorado where erosion has cut down through the loess and into the underlying Ogallala rocks. Some of these canyonlands, like the spectacular Arikaree Breaks in the extreme northwest corner of Kansas and Little Grand Canyon near Wauneta, Nebraska, have scenic values that rival any other place in the Great Plains, but are largely unknown because they occur in remote and thinly populated corners of these states.

You would think that dust is just dust, but geologists have identified several different types of loess in the Great Plains based on the origin of the silt particles and the time of deposition. The loess blanketing western parts of the country of the Kaw is classified as Peoria Loess. Peoria Loess is the most recent of the several episodes of loess deposition in North America and is also called "last-glacial-age" loess because its deposition followed the end of the last episode of continental glaciation. Some of the deepest Peoria Loess deposits in the world are found in the Republican River watershed, in places sixty feet thick. In the canyons of the Arikaree Breaks, where Peoria Loess rests on earlier formations, total loess thickness reaches 180 feet.

Where did all this come from? Loess deposits are usually associated with glaciated areas, where winds have swept up ice-pulverized rock dust from lake and floodplain sediments exposed after glacial retreat. But studies of the mineralogy and geochemistry of Peoria Loess point to siltstones of the White River Group exposed in northwestern Nebraska and southwestern South Dakota. Ancient winds working over these highly erodible rocks liberated silt particles that were whisked away into the country of the Kaw, leaving behind the fantastic badlands of Toadstool Geologic Park in Nebraska and Badlands National Park in South Dakota.

Parent material is the stuff from which soil is formed. Soils often derive from the weathering of underlying bedrock, which strongly influences soil qualities like texture, structure, mineralogy, and fertility. Soils can also form from loess deposits, as has been the case in much of the central part of the country of the Kaw. Two of these loess-derived soil types are of such special significance to agriculture that they are recognized as the official state soil—the Harney Silt Loam of Kansas and the Holdrege Silt Loam of Nebraska.

Areas of sandy habitat occur throughout the Great Plains. A discrete region of sand dunes is called a dune field or, in geologist lingo, a sand sea. The Nebraska Sandhills is the largest dune field in the Great Plains and, at twenty thousand square miles, the grandest sand sea in North America. By comparison, the Wray Dune Field is a "mini-sand sea," but it is the biggest in the country of the Kaw, occupying about twenty-six hundred square miles in northeastern Colorado and southwestern Nebraska, more or less bounded by the valleys of two tributaries of the Republican River, Frenchman Creek on the north and the Arikaree River on the south. Two smaller areas of dunes—the Imperial and Lincoln County dune fields—occur in the Republican watershed to the northeast of the Wray Dune Field and are considered satellites of the Nebraska Sandhills.

The topography of these dune fields varies from low, rolling hills to taller dunes with sharper crests and steeper slopes. Such dunes are said to be "choppy" in the same way that rough seas are choppy, the waves rising and falling in rapid succession. The tallest and choppiest dunes are developed around the town of Wray, Colorado, with some that reach 160 feet in height.

Dunes are a lot like clouds—ephemeral works of wind and weather. And like clouds, if you gaze at dunes long enough recurring shapes and patterns will dawn on you. Geologists have identified and named a number of standard dune forms, seven of which occur in the Nebraska Sandhills. The primary expression in the dune fields of the Kaw is the parabolic dune.

Parabolic dunes have a U or V shape to them, the ridges built up from sand excavated out of a windward blowout. A blowout is a crater-like depression that forms when the cover of vegetation has been thinned and bare sand is exposed to scouring winds. In the case of parabolic dunes, the sand-supplying blowout is located at the open end of the U or V. Parabolic dunes indicate the work of strong, unidirectional winds. The "arms" of the dune point toward the source of the wind, and the "nose" points away. The orientation of parabolic dunes in the Wray Dune Field, their noses aiming southeast, indicates paleowinds out of the northwest.

I never gave sand much thought until I read *The Edge of the Sea* by Rachel Carson. Her description of the origins and ecology of beach sands along the Atlantic Coast is some of the most wonderful nature writing I've ever come across. It turns out that sand grains in Kaw dune fields also have their stories.

Based on geochemistry and relative particle size, it appears that at least some of the sand in the Wray Dune Field came from the South Platte River, which runs about twenty miles to the northwest across a shallow streambed of sand carried down from the Rockies. But evidence also points to the Ogallala Formation as a significant source. Outcroppings of Ogallala bedrock occur upwind from the dune field and, while the upper part of these exposures is capped by erosion-resistant mortar beds, lower portions are uncemented and more easily weathered by the elements, freeing sand grains for the wind to gather up and carry away. Analysis of sands in the Imperial and Lincoln County dune fields indicates a more complicated genesis, with contributions from the South Platte River, the Ogallala Formation, and the Nebraska Sandhills.

The windlands of the High Plains are relatively young in terms of geologic time. Deposition of Peoria Loess in the upper watershed of the Republican River began about twenty-five thousand years ago and ended ten thousand years later. Buried beneath the Peoria are older loess units from earlier episodes of deposition—the Gilman Canyon and Loveland Formations. As for sand dunes, genesis of the present dune fields of the country of the Kaw appears to have begun about three thousand years before present. These dunes, which likewise rest on deeper sands of earlier deposition, were open and actively moving until around fifteen hundred years ago when they were stabilized by the vegetation we see today.

This vegetation, mostly sandsage prairie, is sparse at best and becomes even thinner during periods of persistent drought. Geologists who study the formation of sand dunes and botanists who study the plant communities of dunes warn that these landforms are highly sensitive to climate and always on the brink of reactivation. That makes dune fields in the country of the Kaw important as sentinels of climate change—landscapes we should keep an eye on for early warning signals of tougher times ahead.

The high winds of the High Plains not only shape the contours of the land but can also get into your head. Recent research by anthropologist Alex Velez suggests that the combination of silence and howling wind—the *soundscape* of the Great Plains—was a major catalyst in incidents of depression and even mental breakdown experienced by early Euro-American homesteaders—a malady known during the settlement era as "prairie madness."

Call me crazy, but I'm a fool for the horizontals of the High Plains. One of my favorite drives in the country of the Kaw is along the Hi-Line—Nebraska State Route 23. The highway traverses the southwest corner of the state, with an eastern terminus in the bustling town of Holdrege; from there it heads northwest into Nebraska's rugged Medicine Creek country, climbs up onto an expanse of High Plains flatland known as the Perkins Table, and then exits into Colorado about fifty miles later. This is some of Nebraska's finest wheat country; the tiny village of Venango, the westernmost town on the Hi-Line, proclaims itself "The Buckle of the Wheat Belt." I recommend driving the Hi-Line on an afternoon in June, when you'll have a pretty good chance of catching the spectacle of a purple-green thunderhead boiling up over golden sweeps of ripening wheat.

TWO

Heartland

> The centre of Kansas is very nearly the geographical centre of the United States. It may, therefore, with great propriety be called the CENTRAL STATE. This is well, for it is perhaps the most fertile, as it is unquestionably the most beautiful, of the great sisterhood; and through its gushing heart, as I believe, the great artery of the world's commerce is destined to flow.
> *Josiah Copley,* Kansas and the Country Beyond. . . .

Kansas City boosters have long trumpeted their metropolis as the Heart of America. But the real heart is out in the Smoky Hills.

This melodious name was first applied to a chain of sandstone-capped buttes that stand to the southwest of Salina, Kansas. Rising abruptly from the surrounding plains, these can be seen from a considerable distance as you approach from the east, *smoky* referring to a foggy haze that gathers about them in the early morning. The namesake Smoky Hill River bends around the hills on the south and east as it winds its way to confluence with the Republican.

Smoky Hills is also used as a regional name, although definitions of its geographical extent have varied over the years. The delineation we'll use here is that of the Kansas Geological Survey: the Smoky Hills physiographic region consists of three subsections, each dominated by a successively higher and younger group of erosion-resistant sedimentary rocks—the Dakota Formation on the east, Greenhorn Limestone in the middle, and Niobrara Formation on the west. Others have used the term more narrowly, for the Dakota subsection alone or for a region consisting of the Dakota and Greenhorn subsections. The Smoky Hills region is almost entirely limited to the middle of the Kaw watershed, ranging through north-central Kansas and into a small area of south-central Nebraska.

The rock systems that define the Smoky Hills were formed during the Cretaceous period (roughly 66 to 145 million years ago), when ocean waters covered the interior of the North American continent. The Western Interior Seaway stretched north to south from the Arctic Ocean to the Gulf of Mexico. The width and depth of this epicontinental sea varied significantly over the thirty million years of the Cretaceous period as global ocean levels rose and fell, causing the seaway to expand and "transgress" more of the interior or shrink and "regress." The characteristics of the Dakota, Greenhorn, and Niobrara rocks, and their influence on the modern landscape, reflect where they were deposited in relation to the shoreline and the open waters of the seaway.

Geodesy is the science of measuring the Earth's geometric shape, orientation in space, and gravity. This work, critically important to mapping, navigation, and military intelligence, is carried out under the auspices of the National Oceanic and Atmospheric Administration and uses space-based tools like GPS to generate the needed data. But before satellites, it depended on a network of carefully surveyed, Earth-based triangulation stations or datum points. In 1891, the US Coast and Geodetic Survey set the epicenter of this entire network, the Geodetic Datum for the United States, in a pasture in Osborne County, Kansas.

While the geodetic center was "of utmost practical and scientific value" to the nation, its presence was not of great interest to the general public. It proved easier to draw tourists to another Smoky Hills landmark, the Geographical Center of the United States.

The geographical center of anything is defined as that point on which the surface of an area would balance if it were a plane of uniform thickness. So, if you dug up the United States of America, excluding Alaska and Hawaii, as one big slab, the point at which the slab would perfectly balance is located one mile north and one mile west of the town of Lebanon, Kansas.

This distinction, established in 1898 by the US Geological Survey, had long been appropriated as part of the local identity, so when in 1940 a rogue mapmaker suggested the center might actually be in southern Nebraska, citizens of Lebanon sprang into action and formed the US Hub Club to defend their title.

The club's most visible effort was to erect a monument on the exact spot of the geographic center. Dedicated in 1941, the monument did indeed become a tourist attraction, even leading to the establishment of the U.S. Geographical Center Motel-Café and the U.S. Center Chapel, a one-room roadside wedding chapel.

The tiny, eight-by-twelve-foot chapel looks like a backyard tool shed with a steeple, and its eight little pews can seat only eight adults. Situated in a well-maintained hilltop park, the U.S. Center Chapel is always open, never locked. The original, designed and built in 1967 by Lawrence Tanis, a kindhearted Smith County, Kansas, farmer, was placed in a park in the village of Downs as a wayside sanctuary for weary travelers. A few years later it was purchased by a resident of Lebanon and relocated to the site of the US Geographical Center monument. The original was destroyed in 2008 when a speeding truck missed a turn and crashed into the building, but a replica was soon built to replace it. This charming Smoky Hills attraction was the setting for a poignant 2021 Super Bowl commercial featuring rockstar Bruce Springsteen that sought (in addition to selling Jeep vehicles) to remind a fractured American populace that "we need the middle."

Alaska was not part of the union when the Hub Club built their monument and, to their alarm, the addition of this gigantic state to the slab in 1959 shifted the national center to some place in South Dakota. They eventually rebranded their local attraction as the center of the *contiguous* United States, the lower forty-eight. The Smoky Hills region, the middle country of the Kaw, is still the genuine heartland of America.

The Dakota Hills is the easternmost of the three subsections of the Smoky Hills and is shaped by the attributes of the underlying rocks of the Dakota Formation. These rocks were formed from river-transported sediments deposited in beaches and deltas along the eastern coast of the seaway during its first transgression into the interior of the continent.

Claystones and siltstones occur in the Dakota Formation, the former yielding clays used to manufacture high quality architectural brick at plants in Endicott, Nebraska, and Concordia, Kansas.

But it is the thick Dakota sandstone that gives this region its strong visual character. These sandstones are relatively hard and resistant to erosion, especially exposures where the sand grains are cemented by iron oxide. The Dakota is a handsome rock with colors that range from tan to rust to coffee brown, often in the same outcrop; sometimes it is stained with a black varnish or splashed with bright green patches of lichens.

Most of the Dakota Hills region lies in Kansas, with a small extension into Nebraska along the Little Blue River and its tributaries between the towns of Steele City and Fairbury. While there is a linear regularity and angularity to limestone scarps and hills in the Osage Cuestas and the Flint Hills, the landscape of the Dakota Hills is much more jumbled and disorderly, and the hills more rounded.

Erosion of the Dakota Formation has in places left isolated elevations like Coronado Heights near Lindsborg, which stands about three hundred feet above the valley of the Smoky Hill River. One of the buttes for which the Smoky Hills were named, this landmark is reputed (by local boosters) to have been scaled by Francisco Vasquez de Coronado in 1541 to survey the realm of Quivira. You can scan the scene for yourself from the observation deck of a two-story shelter house that sits atop the butte. Resembling the castle of a Scottish lord, the structure was built in 1936 with support from the Depression-era Works Progress Administration, using Dakota sandstone and a dose of architectural whimsy.

Weathering has also yielded breaks and canyons of varying size and ruggedness, most dramatically where the Saline and the Smoky Hill Rivers have carved into these rocks. The most impressive of these Dakota canyonlands are now occupied by reservoirs, Wilson Lake on the Saline and Kanopolis Lake on the Smoky Hill. A small but scenic cluster of draws and ravines in the Dakota outcrop area of southeastern Nebraska is known locally as Steele City Canyon.

The Dakota Formation is an important aquifer in the Smoky Hills, and springs and seeps occur where groundwater-bearing rock units are exposed along creeks and streams. When groundwater passes through salt-impregnated shales in the Dakota it carries salinity into streams and wetlands in the region. The Saline River and the city of Salina bear witness to this chemistry, as do numerous salt marshes along the eastern edge of the Dakota Hills, the largest being the Jamestown Marsh near Concordia.

Evaporation of water in the shallow parts of these marshes leaves a crust of crystallized salt on the surface; in the past, bison, pronghorn, elk, and deer came to lick the ground to obtain the sodium needed in their diets. These salt licks and salt flats were of such renown they showed up on maps generated from the Lewis and Clark Expedition of 1804–1806, although the expedition never got closer to the Smoky Hills than present-day Kansas City. Many of the salt marshes in the Dakota Hills have dried up or have been degraded by silt washing in from surrounding agricultural lands, but restoration efforts are under way in several places.

In addition to buttes, canyons, and salt, the Dakota is responsible for some notable geological weirdness. Just south of Minneapolis, Kansas, is a remarkable place called Rock City, where over two hundred giant sandstone orbs are scattered across a two-acre park. These "cannonball" concretions formed when compressed sand deposits were impregnated by groundwater with a limy cement that spread from an area of concentration outward to the margins of a sphere. As the surrounding, softer sandstone weathered away, the spheres were left standing on the ground plain. Some of these orbs are over twenty feet in diameter.

The gang at Rock City, designated a National Natural Landmark in 1976, is said to be the largest congregation of sandstone concretions in the world. About twenty miles to the southwest in Ellsworth County is Mushroom Rock State Park, where similar but elliptically shaped concretions are balanced on pedestals of softer sandstone that has not yet fully eroded away.

For untold numbers of years, Dakota sandstone has served as a canvas for Native American rock art. Petroglyphs have been found throughout much of the Dakota Hills in Kansas, particularly in Ellsworth and Russell Counties. The complexity of these inscriptions ranges from simple geometric designs to portrayals of humans on horseback. Sadly, natural weathering of the soft sandstone, coupled with modern vandalism, has defaced and in some cases completely erased these mysterious works of art. At least we have a photographic record of most of the carvings thanks to a systematic survey of Kansas rock art sites by the State Historical Society in 1979 and 1980 and the recent book *Petroglyphs of the Kansas Smoky*

Hills. Clusters of petroglyphs are often associated with springs and seeps at the base of sheltering bluffs, the Dakota providing both a medium and an environment for artistic expression.

The western limit of the Dakota outcrop zone is marked by a northeast-to-southwest trending escarpment formed by the Greenhorn Limestone. This region of rocky, flat-topped hills and benches extends westward to another prominent escarpment capped by the Fort Hays Limestone. Sometimes called the Blue Hills, the country between these two escarpments is the middle range of the Smoky Hills region.

Greenhorn Limestone, the dominant surface rock of the district, consists of thin, chalky limestone beds that alternate with thicker beds of shale. The regional name "Blue Hills" likely owes its origin to outcroppings of dark gray Carlisle Shale that are exposed along creeks and streams between the Greenhorn and Fort Hays escarpments, giving a bluish haze to distant vistas.

The Greenhorn Limestone was deposited on the broad, flat eastern shelf area of the Western Interior Seaway during an episode of maximum transgression. Greenhorn sediments were deposited in relatively shallow waters, but, because they were far enough out from shore, there was little or no input of materials eroded from land. So, they are almost entirely marine in origin. At the top of the Greenhorn is an eight-to-twelve-inch layer of limestone that defines the area geologically and culturally.

Towns throughout the Greenhorn outcrop zone are graced with buildings constructed of this native rock. But you need to head out into the country to see the region's most inventive stonework.

As Euro-American settlers arrived in north-central Kansas in the 1870s and 1880s, fences were needed to define the boundaries of their farms and pasturelands. Although the recently invented barbed wire was an economical fencing material, posts were needed to support and string the strands, and native timber for wooden posts was scarce. Some resourceful person looked to the rocky hills and decided to try harvesting fence posts out of the ground. They not only succeeded in fencing their properties but also chiseled out a one-of-a-kind regional folk art.

The limestone layer that marks the top of the Greenhorn is fairly uniform in thickness and could be split to yield a four-sided column averaging about eight inches on each side, which could then be cut into lengths five or six feet long. Despite the hard work required to quarry, haul, and set posts that weighed 250 to 450 pounds each, far-reaching ranks of stone pillars, spaced about thirty feet apart, would become the visual trademark of the region. The technical geological name of the rock unit that yielded these posts is the Fencepost Limestone Bed of the Pfeifer Shale Member of the Greenhorn Limestone. The locals call it post rock.

Post Rock Country occupies about three million acres in eighteen Kansas counties. As in the Flint Hills, shallow rocky soils have kept much of the landscape in pasture, but instead of tallgrass the rangeland here is mixed-grass prairie dominated by little bluestem, a mid-height bunchgrass that gives a tweedy look to the land. The hill-and-dale topography is delightful in and of itself, but the runs of stone posts—hand-hewn works of art, no two alike, stitched north to south, east to west, into the grassland fabric—unites nature and culture in a way that is winsome beyond words. *Land of the Post Rock: Its Origins, History, and People*, by Grace Muilenburg and Ada Swineford, is an eloquent and essential guide to this most amiable region.

It is estimated that at one time there was nearly forty thousand miles of post rock fence line in the Greenhorn outcrop zone. There is less of it today; the losses began after World War II, when roadways were widened to improve travel and transport. Stone posts set along the old road bed were removed and replaced with steel or wooden ones used for fencing along the new right-of-way. The trend toward larger farm operations also resulted in the removal of post rock fence lines that once divided properties.

But there are still miles and miles of this singular stone craft on display throughout the Greenhorn outcrop zone, which, coupled with the extensive use of post rock in local architecture, makes it one of the most distinctive cultural landscapes in all of America. The Kansas Post Rock Limestone Coalition is a nonprofit organization dedicated to promoting, preserving, and protecting the region's history, art, and architecture.

Running northeast to southwest for about 170 miles, the Fort Hays escarpment sets a prominent eastern boundary for the third and final section of the Smoky Hills. Fort Hays Limestone, the lowermost rock unit of the Niobrara Formation, contributes a great deal of bold topography to north-central Kansas, most notably in the ramparts of its escarpment but also capping outlying hills and buttes that in many places stand a hundred feet or more above the surrounding country. One of the most conspicuous physiographic boundaries in the country of the Kaw, the Fort Hays escarpment imparts a Flint Hills feel to the land, especially in Jewell and Osborne Counties. Strong bluffs have been cut into the Fort Hays Limestone along the middle portions of the Solomon, Saline, and Smoky Hill Rivers. Exposures are massive, blocky, and yellow tan in color.

As you head west of the Fort Hays outcrop area you begin to glimpse patchy exposures of pinkish-yellow rock, especially in the dissected hills along major streams in the region. This is Smoky Hill Chalk, the younger, upper member of the Niobrara Formation. The rock units of the Niobrara Formation were laid down during the latter part of the Cretaceous period, when the Western Interior Seaway was at one of its maximum transgressions. Fort Hays sediments were deposited in relatively deep waters far from shore where there was little or no input of materials eroded from land and so are almost entirely marine in origin. Smoky Hill Chalk was deposited later when the sea was still wide and deep but was regressing from its maximum invasion.

Chalk is a soft, fine-textured limestone that forms in an open ocean environment; it is composed mainly of calcium carbonate from microscopic marine plants and animals that floated or swam near the surface. Pure chalk is white, but iron oxide and other impurities in Smoky Hill Chalk often give it a tinge of pink.

Farther west along the Smoky Hill River and its tributaries, erosion has carved more deeply into the thick chalk deposits, sculpting rugged canyons, stark badlands, and an assortment of spires, hoodoos, and columns that stand as outliers away from the bluff-lined walls of the valley. The more spectacular monoliths have their own names—Castle Rock, Monument Rocks, and the Pyramids—as do the more striking stretches of badlands like Goblin Hollow, Hell's Half-Acre, and Little Jerusalem.

These erosional remnants became famous landmarks after travel started along the Smoky Hill Trail in the 1850s. The trail, which followed the Smoky Hill River across central and western Kansas, offered a shorter, if riskier, route to the Colorado gold fields than either the Platte or Arkansas Rivers. Now iconic images of Kansas calendars and coffee table books, these geological oddities are located on private land, but access to the public is generally granted. For an immersion experience in chalk geology, the recently opened Little Jerusalem Badlands State Park on the Smoky Hill River south of Oakley offers hiking trails to take you into 220 acres of badland terrain.

Smoky Hill Chalk crops out in a limited area of south-central Nebraska and is exposed in the dissected hills along the Republican River from the town of Superior west to the Harlan County Reservoir. While less dramatic than the carvings along the Smoky Hill River, these outcrops still impart a strangeness and beauty to the land.

A few miles upstream from Superior is the town of Guide Rock, named for a nearby hill on the south side of the Republican River that was part of the sacred geography of the Pawnee people—*Pa:hu:ru'*, "the rock that points the way." A 1904 photograph shows the hill to be a striking landmark, with an upward-angling prow and perpendicular face of Smoky Hill Chalk bedrock. In Pawnee mythology there are certain places where animals with miraculous powers would gather to hold councils. Pa:hu:ru' was one of five such "animal lodges" in present-day Nebraska and Kansas where the *Nahu'rac* met belowground to confer supernatural power on certain worthy human beings.

The Guide Rock was also a significant milestone along the storied Pawnee Trail, the route entire Pawnee villages would follow twice a year between their earth lodge towns on the Platte River and the rich bison pastures of the country of the Kaw. Sadly, this historically and culturally important site is now so densely covered with trees, particularly eastern red cedar, that it is hard to make out its shape and distinguish it from other points in the bluff line.

Surely Willa Cather was remembering Republican River chalk bluffs when she wrote "The Enchanted Bluff." The celebrated author of *O Pioneers* and *My Ántonia* grew up in Red Cloud, in sight of the Republican valley, and published her short story in *Harper's Monthly Magazine* in 1909, several years before her first novel was in print.

The story opens with six Nebraska country boys relaxing after an end-of-summer swim, cooking supper over a driftwood fire on an island in a river bordered on one side by "an irregular line of bald clay bluffs." Sitting around the campfire, looking up at the stars, they were startled by a very sudden moonrise: "We all jumped up to greet it as it swam over the bluffs behind us. It came up like a galleon in full sail; an enormous, barbaric thing, red as an angry heathen god."

The breathtaking spectacle provoked the boys to dream out loud about wondrous places they would visit when they got older. One of them, Tip Smith the grocer's son, had heard of a place in New Mexico called Enchanted Bluff—a lone red rock no white man had ever been to the top of. His description so stirred the lads that each vowed to one day find it and climb it. But their youthful aspirations, sparked by the sight of that moon over those Republican River bluffs, got lost in the tangles of adulthood.

Birger Sandzén never intended to stay in Kansas. The budding Swedish artist arrived in the town of Lindsborg in the fall of 1894 after he had spent several months studying in Paris. He planned to work for a short time at the recently founded Bethany Academy, get a feel for the American landscape, and then head back to Europe to pursue his painting in earnest. But in Lindsborg he met the lovely Alfrida Leksell, and the picturesque Smoky Hills. The young man's heart was captured by the both of them, and his three-year sojourn stretched into sixty.

Sandzén taught art and languages at the academy, later named Bethany College, from 1894 until his retirement in 1946. A prolific artist, he produced an estimated twenty-eight hundred oil paintings, four hundred watercolors, and over five thousand sketches. His unique brush techniques and vivid color palette gave his oil paintings a stunning luminosity, and earned him acclaim as one of the premier American landscape artists of the time.

A wonderful collection of his paintings and other work can be viewed in the Birger Sandzén Memorial Gallery on the campus of Bethany College. Or you can visit the post offices in Lindsborg or Belleville, Kansas, each of which has a large Sandzén mural on display. Commissioned under the auspices of the US Treasury Department's Section of Fine Arts, a Depression-era program designed to support artists by funding paintings and sculpture for federal buildings, both murals depict river valley landscapes in Sandzén's vibrant yet pastoral style, *Smoky River* (1938) in Lindsborg and *Kansas Stream* (1939) in Belleville.

Much of Sandzén's work portrays the Dakota Hills countryside around Lindsborg. He had a special fondness for the warm coloration of the Dakota sandstone, and would take his students to Coronado Heights and Salemsborg Hill north of Lindsborg to sketch and paint. But he also depicted farmsteads with houses and outbuildings built of limestone, probably Fencepost or Fort Hays. And, because Alfrida's folks had a farm out west near Hill City, Kansas, he spent many hours sketching and painting the landscape traversed by Wildhorse Creek, a tributary of the South Fork of the Solomon River, where chalk of the Niobrara Formation is exposed. His Nordic soul seems to have been drawn to rocky places. Born in Sweden in 1871, he died eighty-three years later, a true son of the Smoky Hills.

Though based in Lindsborg, Sandzén traveled widely in the American West, painting powerful scenes from places like the Colorado Rockies, Yellowstone, the Grand Canyon, and the California coast. But his heart was firmly rooted in Kansas. In a 1915 article in the *Fine Arts Journal*, Sandzén described the wave of eastern artists discovering the visual drama of the West, with some even starting to settle in places like Taos, New Mexico. He understood the appeal of "romantic and spectacular" scenery, but was staying put: "There will be no 'artists' colonies' springing up at the foot of the quiet 'Smoky Hills' in Central Kansas or on the 'rolling prairie' . . . and, to be perfectly frank about it, we few painters out here feel quite grateful for being left alone with our humble friends."

THREE

Scarplands

> Early writers in Kansas referred to these hills as the Kansas Mountains.
> *Nathan Wood Bass*, Geology of Cowley County, Kansas

If you could dye the waters of the Kaw and its entire network of tributaries fluorescent orange, and get someone on the International Space Station to snap a picture for you, you would see that the river system sits squarely in the middle of the North American continent, about equidistant from the Pacific Ocean on the left and the Atlantic on the right. This might give you a settled sense of the river's place on the planet, but your vision would be a bit shortsighted. The rocks tell a much more staggering story.

Some three hundred million years ago, North America was in a tightly fitted relationship to the Earth's other land masses in a giant supercontinent called Pangaea. The land that is today the country of the Kaw was situated on the northwest margin of Pangaea, in a coastal area on the continental shelf. Collision of tectonic plates within Pangaea gave rise to the ancient Appalachian and Ouachita mountain systems, and rivers coming down from these highlands carried enormous burdens of eroded sediments to the coast that accumulated in deltas, alluvial plains, and tidal flats. Meanwhile, massive ice sheets generated out of the southern polar region of the supercontinent advanced and retreated, resulting in global fluctuations in sea level.

The combined effects of these titanic processes on the coastal areas of Pangaea were episodic alteration between marine and terrestrial environments and a back-and-forth migration of the shoreline. Over immense stretches of time, a recurring sequence of rock layers was laid down: limestones in the warm, shallow seas; shales and sandstones in deltas, swamps, and estuaries along the shore.

The fantastic story of Pangaea is revealed in rocks exposed in the eastern reaches of the Kaw. The older of these are of the Pennsylvanian period (299–323 million years ago) and are found between Kansas City and the Flint Hills in a region called the Osage Cuestas. The younger are of the Permian period (252–299 million years ago) and are the stuff out of which the Flint Hills were built. The unifying characteristic of these two regions is a series of east-facing, northeast-to-southwest trending escarpments called cuestas.

Cuestas develop where alternating strata of hard and soft sedimentary rocks are slightly tilted from the horizontal, resulting in a steep, cliff-like face on one side and a gentler dip slope on the other. Cuestas in our region dip to the west and northwest, due to the uplift of the ancient Ozark Mountains region to the southeast. The crest of each escarpment is capped by an erosion-resistant limestone, while the dip slope is developed in weaker shales or sandstones. These are the scarplands of the country of the Kaw.

Stand on the bluffs of Strawberry Hill, on the Kansas side of Kansas City, look across the valley of the Kaw to the Missouri side, and you can spot Quality Hill, just about a mile away. Strawberry Hill was settled in the late 1800s by European immigrants who came to this river city to work in the stockyards, meatpacking plants, rail yards, and other hard-labor industries that covered the valley below—an area known as the West Bottoms. A mix of Croatians, Serbs, Italians, Irish, Germans, Russians, and others, they built modest homes and beautiful churches up on this high ground to escape the flooding that often plagued the bottomlands of the Kaw. Mexican Americans would later add to the rich cultural fabric. Strawberry Hill still has the look and feel of an ethnic neighborhood.

The quaint name is said to have been inspired by the abundance of wild strawberry plants growing on this upland. The origin of "Quality Hill" is not so heartwarming. This unabashedly pretentious branding was given to Kansas City's first upscale planned neighborhood, established in 1857 by financier Kersey Coates. In its heyday, Quality Hill was the most fashionable and pricey enclave in Kansas City, a mansion district full of wealthy merchants, bankers, captains of industry, politicians, and other power brokers.

Quality Hill eventually lost its luster, in part because of the smoke and odors that wafted up from burgeoning industries down in the bottoms, but especially after squatters began to build shanties on the lower slopes of the bluffs. The area underwent significant decline, but has made a comeback in recent years as a trendy historic neighborhood.

Although these two elevated landmarks initially attracted folks from very different socioeconomic strata, Strawberry Hill and Quality Hill share a common geological underpinning—the Argentine Limestone. Named after the Argentine district of Kansas City, Kansas, this Pennsylvanian rock layer dominates much of the surface geology of older parts of the metro area, often forming prominent cliffs and bluffs.

Standing about two hundred feet above the West Bottoms, both hills offer fine views of Kaw Point, the confluence of the Kaw with the Missouri. The vista encompasses a riverine landscape that figured prominently in the early history of Kansas City, and in the exploration of the American West. The Lewis and Clark Expedition camped at Kaw Point on June 26, 1804, on the westbound leg of their transcontinental Journey of Discovery. Then, on September 15, 1806, on their way back to St. Louis, the two cocaptains climbed to the top of what would later be known as Quality Hill because, as William Clark wrote in his journal, it "appeared to have a Commanding Situation for a fort."

Another prominent layer of Pennsylvanian rock in the Kansas City area is the Bethany Falls Limestone, which forms the picturesque bluffs and glades found in Swope Park and at other places along the Blue River on the south side of the metro, giving the upper Blue a rocky bottom and shoals like an Ozark stream. When I was a boy, my buddies and I would hike to an area along the Blue where large blocks of the Bethany Falls had broken away from the main outcrop, creating a warren of narrow passages we dubbed "Indian Hideout."

But of grander consequence is a ledge of Bethany Falls Limestone that crops out at water's edge along the south bank of the Missouri River, which during the riverboat era provided an excellent place to land and unload cargo. The proximity of this natural rock wharf at the elbow bend of the Big Muddy, dubbed Westport Landing, precipitated a shift of overland trail traffic from Independence to Westport and led to the founding of the Town of Kansas, Kansas City's precursor, in 1839.

Head west from the jumbled terrain of downtown Kansas City, away from the disorienting tangle of bridges and viaducts that crisscross the industrialized final few miles of the Kaw, and you will encounter the calming series of northeast-to-southwest trending ridges that define the Osage Cuestas region. In the words of John Charlton of the Kansas Geological Survey, these cuestas "extend across the prairies of eastern Kansas like waves."

The Argentine and Bethany Falls Limestones, so evident around Kansas City, are part of this rhythmic geology, but the boldest cuesta in eastern Kansas is formed by the Oread Limestone, which runs through the vicinity of Lawrence. The Oread is the substance of Mount Oread, the hilltop on which the University of Kansas stands.

The topography of the Osage Cuestas provided a perfect theater for the Border War, a bloody period of depredation and reprisal by bands of proslavery marauders and abolitionist vigilantes that began on the Kansas-Missouri border before the Civil War and lasted nearly two decades. The forest-cloaked, ridge-and-vale terrain favored the hit-and-run guerilla-style raids that characterized the conflict.

But the geology and ecology aided defense as well as offense. So frequent were the sorties of proslavery bushwhackers out of Missouri that an early-warning system was devised using a towering white oak tree on a high ridge about fifteen miles south of the abolitionist stronghold of Lawrence. Upon the sighting of advancing raiders, signals were hung in the branches of the oak—flags by day and lanterns by night—that could be seen from Blue Mound, an isolated erosional remnant of Oread Limestone to the north, and from there relayed to sentries on Mount Oread. The Signal Oak died in 1914, and a historical marker was placed on the ridge where it silently carried out its duties, about two miles north of Baldwin City, Kansas.

The Flint Hills stretches north to south along the western edge of the Osage Cuestas, from near the Nebraska border down into northern Oklahoma. The boundary between the two regions is marked by a prominent east-facing escarpment of Permian bedrock fronted by a series of terraced rocky slopes. Once you've worked your way up through these step-like benches and reached the rolling Flint Hills upland, you've gained about 350 feet in elevation.

Like an elm leaf in outline, the Flint Hills region is broadest just above the middle, which is about where the Kaw forms from the union of the Republican and Smoky Hill Rivers. The Kaw and its tributaries drain most of the northern third of the Flint Hills. Some of the same Permian rock formations that define the Flint Hills extend into southeastern Nebraska, where they are exposed in the hilly terrain of the Big Blue River country south of Beatrice. While not considered part of the Flint Hills, this area has its own distinctive rockiness.

As in the Osage Cuestas, the topography of the Flint Hills is controlled by a series of limestones that are harder and more resistant to erosion than the shales with which they are interbedded. Among the more prominent are the Cottonwood, Florence, and Fort Riley Limestones. But an important difference is that the scarps and dip slopes of Flint Hills cuestas are more closely spaced, resulting in a more rugged landscape.

The other major difference is that the limestones of the Flint Hills contain numerous bands of chert. Commonly called flint, chert is a very fine-grained sedimentary rock made up of interlocking crystals of quartz; it is harder than limestone and practically insoluble. Chert-bearing limestone beds are more resistant to weathering, but also, when chert does eventually break down, it liberates a flinty gravel that accumulates on the surface of the land. This chert armor further slows erosion and preserves the elevated ridges and benches that give the Flint Hills its striking visual character.

If not for the flint of the Flint Hills, there would be very little tallgrass prairie left in America. The region lies at the western edge of the original presettlement distribution of the mid-continental tallgrass prairie. Due to the rockiness and ruggedness of the land, much of it is unsuited to cultivated agriculture so, while tens of millions of acres of tallgrass prairie were plowed under throughout the Midwest in the 1800s, the Flint Hills was mostly left in grass. Thanks to its cherty limestones, over four million acres of tallgrass prairie, one of the most diminished and imperiled ecosystems on Earth, is preserved in the Flint Hills. This geology also yields rocky watercourses with relatively clear-running streams that support assemblages of native fish populations not found in the muddied waters of agricultural areas.

Unfold a road map for the state of Kansas, and the outline of the Flint Hills will be evident by the greater density of dots and squiggles on either side of the region. This is cowboy and cowgirl country, some of the finest grazing lands in the United States, and towns and roads are fewer and more dispersed than in the farming regions to the east and the west. An island of elbow room set between corn and wheat country.

Old-timers called this region the "Kansas Mountains." You might be tempted to dismiss this name as the christening of an enthusiastic but untraveled yokel, who had never seen *real* mountains, or maybe just snarky hyperbole. But find your way to one of the highest points in the Flint Hills, like the uplands east of Cassoday where the Cottonwood, Fall, and Verdigris Rivers share headwaters, and you'll think you're on top of the world.

It was on this highland that Zebulon Pike traversed "very ruff flint hills" on September 10, 1806, and his journal entry for the day provided inspiration for the region's beloved toponym. The scene he witnessed was like flying over the Serengeti: "I stood on a hill, and in one view below me saw buffalo, elk, cabrie [pronghorn], and panthers."

While we can't trace the exact route of Pike's journey across the Flint Hills, the state of Kansas has designated a number of scenic drives through the region that offer similarly wonderful vistas. The Native Stone Scenic Byway and Skyline/Mill Creek Scenic Drive traverse very picturesque stretches of Flint Hills terrain within the country of the Kaw.

The combination of rugged topography, horizon-to-horizon sweeps of tallgrass, and immense skies makes the Flint Hills one of the most stirring landscapes on this continent. I once had the pleasure of leading a tour of the Flint Hills for a small group that included famed English horticulturalist Roy Lancaster. We were hiking along a trail on the Tallgrass Prairie National Preserve near Strong City when I realized that Roy, striding on ahead of me, was singing. The globe-trotting plant explorer explained that the land reminded him of the heather-cloaked moors of his native Lancashire, where he wandered as a young lad and fell in love with the world of plants. The exuberant Englishman was so overcome he just had to sing.

Felsenmeer, German for "sea of rocks," is not a word you expect to use in Kansas or Nebraska. You might come across it while reading about the ecology of Arctic or alpine regions. Yet certain prairie hilltops in the country of the Kaw are weirdly covered with a dense veneer of cobbles and boulders. Not that rocks are unexpected, but these rocks aren't locals. They came from somewhere else.

North America underwent four major periods of continental glaciation, when ice sheets originating in the polar Arctic grew and spread southward into what is today the north-central and northeastern United States. The earliest of these "ice ages," previously known as the Kansan but now called the pre-Illinoian, had the most wide-spreading ice sheet, which flowed all the way into the scarplands of the Kaw.

We know this because of the distribution of glacial erratics in the region. An erratic is a rock fragment carried by glacial ice, deposited at some distance from the outcrop from which it was derived. Also called "dropstones," erratics found in our region are mostly fragments of Sioux quartzite, a rock formation that crops out in southwestern Minnesota and adjacent South Dakota. As the ice sheet flowed across surface exposures of the Sioux quartzite, it cracked and fractured the bedrock, breaking off chunks that became locked in the ice and were carried along. When the ice reached its southern limits and eventually began to melt and retreat back north, its load of rock was dropped along the terminus, leaving a mystery for us to ponder a few hundred thousand years later.

The footprint of glacial maximum in the country of the Kaw can be traced by the distribution of Sioux quartzite erratics. This very distinctive rock is dense, heavy, and rose purple in color. The erratics range in size from pebbles to cobbles to boulders. Within the glacial drop zone, you may encounter uplands crowded with modestly sized chunks or a lonely block the size of a refrigerator. Either way, the sight is a bit jarring, especially when you consider the rocks traveled some three hundred miles to get here.

Erratics are one component of what is called glacial drift or glacial till, a mix of clay, silt, sand, gravel, and rock chunks that glaciers dragged along and then deposited on the landscape as the ice melted. The overall area impacted by glaciation in eastern Nebraska and northeastern Kansas has been called the Dissected Till Plains. It covers the northern reaches of rock types that define the Osage Cuestas and Flint Hills, plus the very northeast corner of the Smoky Hills.

The Delaware River and Stranger Creek, south-flowing tributaries of the Kaw, are the main stream systems that drain the heart of the glaciated region, which we'll call the Glacial Hills. The terrain is a gently rolling plain and, because glacial drift weathers into very good farm ground, it is a mostly agricultural landscape.

When you compare the distribution pattern of glacial erratics with stream courses in the region, it is evident that the long-gone ice also profoundly shaped the land. From such clues it appears that the Big Blue River and its tributaries were once ice-margin streams, since their generally northwest-to-southeast trends run along what the erratics indicate would been the west-facing edge of the ice sheet.

Located in the valley of the Big Blue River, the town of Blue Rapids, Kansas, has embraced the glaciated past of its setting with a one-of-a-kind Ice Age Monument located in the center of its one-of-a-kind *round* town square. The monument features excellent interpretive signage and an assortment of very large glacial erratics hauled in from the surrounding countryside.

The course of the Kaw itself, from near Wamego to its junction with the Missouri River, appears to have been set by the southern edge of the ice sheet. Glaciation even favored the genesis of Kansas City, giving rise to the celebrated elbow bend of the Big Muddy where it abruptly changes from a south-flowing stream to an east-flowing one, an arc that made the environs of Kaw Point a natural hub for trade and transport.

Some of these glacier-lugged rocks have continued their travels via human agency to city parks, courthouse lawns, museum grounds, and other domestic settings. The most poignant resting place I've come across is in the cemetery at Red Cloud, Nebraska, where a glacial boulder the size and shape of a grand champion watermelon marks the graves of Marian and Bertrand Schultz.

Bertrand, who was born and raised in Red Cloud, was a distinguished geologist and paleontologist and was director of the University of Nebraska State Museum for thirty-five years. He and Marian shared a deep love of Nature, adventure, and travel. They acquired the boulder, a uniquely patterned chunk of conglomerate rock, from a farm near Lincoln and displayed it on their property on the north edge of town. After Marian's passing in 1992, Bertrand decided the stone should be used to mark their shared gravesite, set on a granite base engraved with the words,

"SEARCHING AROUND THE WORLD FOR GOD'S WONDERFUL CREATIONS." He joined her in 1995.

But the most famous glacial erratic in our region is the Shunganunga boulder. In 1929, this huge chunk of Sioux quartzite, ten feet on edge and estimated to weigh twenty-eight tons, was relocated from the banks of Shunganunga Creek near Topeka to a park in Lawrence and made into a monument for the city's seventy-fifth anniversary celebration. The monolith was extracted from the creek bed by crane, transported to Lawrence by railcar, hauled to the park, set on end, and fitted with a plaque honoring the city's founders.

Civic leaders in Topeka were incensed when they learned of the heist, especially since a campaign was under way to move the rock to the grounds of the state capitol. But the question of which town had claim to the boulder was complicated by the fact that it is sacred to the Kansa people, who would visit *Iⁿ'zhúje'waxóbe* (pronounced EE(n) ZHOO-jay wah-HO-bay), the Sacred Red Rock, at the confluence of Shunganunga Creek with the Kaw and there pray and make offerings.

In 2020, the Kaw Nation asked the city of Lawrence to return the rock to the tribe, and in March 2021 the Lawrence City Commission voted to do so. In the summer of 2023, the Sacred Red Rock was moved to the Allegawaho Memorial Heritage Park, land owned by the Kaw Nation near Council Grove, Kansas, about sixty miles to the southwest.

Your best chance of running into a prairie felsenmeer in our region is to go for a drive in the hills around Wamego, Kansas, or Fairbury, Nebraska. The proximity of a sea of glacial erratics can also be detected by driving around in town, where you'll find Sioux quartzite stones enthusiastically incorporated into all manner of local architecture. Wamego has the iconic fountain and bandstand in City Park, setting for the annual Tulip Festival. Fairbury has the enchanting Boy Scout cabin in McNish Park. The storybook English Tudor–style cottage, reputedly built by men employed by the Works Progress Administration during the Great Depression, looks suspiciously like the work of elves.

We'll come across many examples of the bond between the lithosphere and biosphere as we continue our exploration of the country of the Kaw, but we can't close out our three-chapter dive into geology without a brief consideration of fossils.

Charles Sternberg probably thought a bit too much about paleontology—it even invaded his sleep. In his memoir, *The Life of a Fossil Hunter*, the renowned Kansas fossil collector recounts a dream in which, as he explored the hills along the Smoky Hill River south of Ellsworth, he stumbled across a treasure trove of beautiful fossil leaf imprints. Sternberg had already spent time searching the area and had found many good fossils in his wanderings, particularly in a place he named Sassafras Hollow. The setting of his vision was a different locality but the details of the scene were so vivid he felt compelled to go look for it. Guided by dreamscape landmarks, Sternberg "went to the place and found everything just as it had been in my dream."

Some of the most important and spectacular fossil discoveries known to science have been made in the country of the Kaw, many of them by Charles Sternberg and his son George. If this sounds like regional boosterism, let me present a sampling of evidence.

One of the world's richest deposits of insect fossils was discovered in Permian rocks near the Flint Hills town of Elmo, Kansas. Over 150 insect species have been described from these fossils, including a gigantic dragonfly-like creature with a wingspan of twenty-nine inches. Given the name *Meganeuropsis permiana*, the insect, the largest ever known to exist, cruised a swampy, forested lowland area around a freshwater lake.

The earliest-known fossil flowers were discovered in association with the Dakota Formation in the Rose Creek watershed of south-central Nebraska. Initially dubbed the "Rose Creek flower," the plants that produced the five-petaled flowers were later given the lyrical genus name *Dakotanthus*—"Dakota-flower." These early flowering plants occurred in coastal areas on the eastern margin of the Western Interior Seaway of the Cretaceous period in forests that hosted a rich diversity of tree species. Leaves dropped from these trees left exquisitely detailed impressions in the sandstones and mudstones of the Dakota, like those Charles saw in his sleep.

One of the most famous fossils in the world is the "fish-within-a-fish" displayed in the Sternberg Museum of Natural History in Hays. Collected by George Sternberg from Smoky Hill Chalk badlands, the stunning specimen, dubbed "the impossible fossil," shows a six-foot-long *Gillicus* fish inside the rib cage of a nearly fourteen-foot-long *Xiphactinus* fish. These fishes and others shared the Cretaceous seas with mosasaurs and plesiosaurs—giant marine reptiles with elongate bodies and paddle-shaped limbs for swimming. Specimens and replicas of these fearsome creatures from the Smoky Hill Chalk grace museums all over the world, including a forty-foot mosasaur in the KU Natural History Museum.

The Smoky Hill Chalk has also yielded some of the world's finest fossils of prehistoric flying creatures. Pterosaurs were flying reptiles that glided over the Cretaceous sea on wings spanning up to twenty-four feet across. Material collected by George Sternberg became the basis for the description of a new species (initially named *Pteranodon sternbergi*, then *Geosternbergia sternbergi*) with a huge cranial crest that projected upward and backward from the skull like that of a giant maniacal blue jay. Fossils of the primitive bird, *Hesperornis regalis*, were first discovered in association with Smoky Hill Chalk in western Kansas. This strange, flightless marine bird lived most of its life in the water and preyed on small fish.

The largest mammoth skeleton on display in the world is the star attraction and mascot of the University of Nebraska State Museum. The enormous Columbian mammoth, fourteen feet tall at the shoulder, was discovered on a ranch in the Loess Canyons region of southwest Nebraska. Its nickname, "Archie," was derived from an early scientific name for the Imperial mammoth, *Archidiskodon imperator*. Archie was a creature of the much more recent Pleistocene epoch, a time of repeated cycles of continental glaciation when the overall climate was much colder and drier than today.

In his national bestseller *A Short History of Progress*, Ronald Wright describes the frightful unwinding of some of the world's mightiest cultures, from Ur in Mesopotamia to the Maya of Mesoamerica to the Roman Empire. The common thread, "after reading the flight recorders in the wreckage of crashed civilizations," is what he calls a "progress trap"—environmental collapse wrought by human ingenuity.

Relocation of the Shunganunga boulder to Allegawaho Memorial Heritage Park was absolutely the right thing to do, culturally and morally. It also had the side benefit of making this magnificent glacial erratic more accessible than it had been in Lawrence's Robinson Park, a small plot of land hemmed in on three sides by very busy streets. For this we can all be thankful, because pilgrimage to the rock could serve as an antidote for the ecological shortsightedness that threatens us yet today.

Geology, like theology, should humble us, maybe even unnerve us. From the High Plains to the Osage Cuestas, from Cedar Point to Strawberry Hill, the country of the Kaw presents a book of rocks, each chapter a story of forces and time frames that overload our mental circuitry and dwarf our most sophisticated technologies and brilliant master plans.

So, load up the gang and go visit *Iⁿ'zhúje'waxóbe*. Consider that an ice sheet, *one mile thick*, busted this thing loose from bedrock and toted it to Kansas, maybe three hundred miles, at a speed of about ten feet a year. That is a trip from my couch to the refrigerator, once a year, for 158,000 years. For perspective, the first human beings to gaze upon the Great Plains, the people of the Clovis culture, walked in from the Pacific Northwest a mere twelve thousand years ago.

Just because we're small potatoes in the grand scheme of things doesn't mean we're not important. But it is in our best interest to be ever mindful of that smallness.

FOUR

Tree Folk

> Trees were so rare in that country, and they had to make
> such a hard fight to grow, that we used to feel anxious
> about them, and visit them as if they were persons.
> *Willa Cather*, My Ántonia

The forests of the Southern Appalachians are the standard by which all others are judged. More species of trees occur in these ancient mountains than in any other forest type in North America, complemented by an equally lavish array of shrubs and wildflowers. Hike into one of the magnificent cove forests of the Great Smoky Mountains of Tennessee and North Carolina, sheltered in a harbor-like mountain fold, and you are standing in one of the most diverse plant communities on Earth.

The Appalachian Highlands anchor the eastern deciduous forest of North America, which extends westward all the way to the confluence of the Kaw with the Missouri. Ecologists recognize several regional variants of the eastern deciduous forest, distinguished by the species of trees that dominate the canopy. If you drive along a transect from the Shenandoah Valley of Virginia west to Kansas City, nearly a thousand miles, you will begin in the mixed mesophytic forest of the Appalachians, cross the western mixed mesophytic forest of the Interior Highlands, and end up in the Ozarkian oak-hickory forest of western Missouri and adjacent Kansas.

Stop the car at points along the way and survey the local flora. You will find the overall diversity of trees, shrubs, wildflowers, ferns, and mosses gradually decreasing as you move from east to west. So, if you're a lover of trilliums you might as well head back to Virginia where there are ten species of this enchanting wildflower, because only one grows in Kansas City area woodlands.

Mesophytes are plants adapted to environments that are neither particularly dry nor particularly wet. The mesophytic forests of the Appalachians are dominated by towering deciduous trees that flourish in relatively sheltered, well-watered environments. Prime examples include American beech, tulip poplar, basswood, sugar maple, and, historically, American chestnut, along with a few evergreen species like Eastern hemlock. West of the mountains these species gradually decrease in importance and, as the regional climate becomes drier, are replaced by oaks, hickories, and other trees that can handle less moisture.

While there is a broad and gradual change in the dominant tree species from the Shenandoah Valley to Kansas City, another transition, much finer in scale, takes place at the western fringe of the eastern deciduous forest. Depending on slope, exposure, soil type, and underlying geology, wooded communities in western Missouri and eastern Kansas grade from forest to woodland to savanna. Ecologists classify a wooded community as forest when the stand of trees is dense enough that the crowns touch and the canopy is mostly or fully closed, sometimes expressed as 61 to 100 percent tree cover. Woodlands are more open, with canopy cover of about 30 to 60 percent. Canopy cover is less than 30 percent in savanna.

Oak-hickory forest in the Kansas City area is primarily associated with rivers and streams. The dominant trees of upland areas vary with slope and exposure: white oak, black oak, shagbark hickory, white ash, ironwood, and redbud are found on drier upper slopes and red oak, basswood, black walnut, and Kentucky coffeetree in more sheltered situations. In bottomland forests, the major players are cottonwood, American elm, hackberry, silver maple, sycamore, green ash, and bur oak.

Differences in canopy cover among forest, woodland, and savanna also translate into differences in the composition of the subcanopy and ground flora. Forests typically have a well-developed, multilayered subcanopy of shade-tolerant trees, shrubs, vines, ferns, and herbs. In contrast, woodland and savanna have a sparse understory of smaller trees and shrubs and a ground flora rich in prairie grasses and wildflowers.

Chestnut oak and post oak are the dominant tree species of open woodland and savanna in the Kansas City area, and chestnut oak prevails on especially rocky sites. A savanna-grown oak has a distinctive architecture—open and sparingly branched, the wide-spreading limbs often heavy and gnarled. Freed from the crowded forest, a savanna oak looks like someone who has just stepped out into the sunshine and is taking in a very deep breath.

You have to use some imagination to see oak savanna today. Historical evidence can be found in land surveys conducted by the US General Land Office; in Jackson County, Missouri, these took place in 1826–1827 just before farms were being established in the Kansas City area. Surveyors noted when they were in prairie and when they were in timber and used the terms "scattered timber" and "barrens" to denote what ecologists would today classify as savanna, of which there was a fair amount.

But savanna is a fire-maintained community, and the suppression of periodic fires that came with settlement allowed encroachment of shrubs and understory trees into the formerly grass-dominated herbaceous layer of oak savanna. Hints of savanna can be seen today in places where old oaks have the spreading, sun-embracing character of an open-grown tree, but the understory beneath them is crowded by the infill of lesser trees and shrubs.

The picture that emerges from the Kansas City–area land surveys is an intergrading mosaic of forest, open woodland, savanna, prairie, and rock outcrop communities called glades. This gradual shift from forest to prairie happens all along the western front of the eastern deciduous forest, from Manitoba south into Texas. The margins were historically mediated by periodic fires, sparked by lightning or set by indigenous peoples; these burned through grasslands with no ill effect on prairie grasses and forbs but were deadly to trees and shrubs that have most of their biomass aboveground.

This ecological transition is also something of a cultural divide. During the 1800s, the Brulé band of the Lakota were among several tribes that hunted bison on the Republican and Smoky Hill Rivers. But before their days as plains horse nomads, the Brulé were part of a larger amalgamation of Siouxian peoples residing in present-day Minnesota, in an ecological tension zone between forest and prairie. Those, like the Brulé, who ventured out onto the buffalo plains referred to kinsmen who lingered in wooded country as *Saone*—"shooters among the trees." It was not meant as a compliment.

Unlike the dynamic give-and-take among forest, woodland, savanna, and prairie in the uplands, the composition of floodplain or bottomland forest in the eastern reaches of the Kaw is fairly uniform and stable. Major species along the Kaw and its larger tributaries are green ash, American elm, hackberry, cottonwood, willows, silver maple, and sycamore.

Yet most floodplain forest species eventually drop out of the mix as you move farther west. Such is the case of the sycamore, iconic river bottom tree of the eastern United States. The moonlight-infused trunks of sycamores light up the valley of the Kaw from Kansas City to Topeka but then become scarce, barely persisting across the Flint Hills and dropping out of the valley of the Big Blue River just south of the Nebraska border.

As the Kaw cuts across the Flint Hills, it occupies a well-defined rock-walled valley from two to three miles wide that supports a rich floodplain forest. But in the more dissected topography north and south of the river valley, trees are almost entirely confined to narrow ravines and draws in what are called gallery forests. Chestnut oaks dominate the upper edges of these woodlands, and bur oak, bitternut hickory, and hackberry prevail on the lower slopes. Redbuds, one of the most drought-tolerant trees in the Flint Hills, often break free and work their way up the grass-covered hillsides, filling rocky draws with a rose-tinted mist in April.

Beyond the Flint Hills, it falls to two canopy makers—bur oak and cottonwood—to carry the flickering torch of the eastern deciduous forest a bit farther up the tributaries of the Kaw. Although occurrences of American elm, green ash, hackberry, black walnut, and a few other easterners can also be found to the west, as attested to by the number of Elm, Ash, Hackberry, and Walnut Creeks in the middle reaches of the Republican and Smoky Hill, bur oak and cottonwood are keystone species of unique Great Plains woodlands to which each imparts a distinctive character and ecology.

Twelve different species of oaks populate the forests and woodlands of the Kansas City area: black, blackjack, bur, chestnut, dwarf chestnut, pin, post, red, shingle, Shumard's, swamp white, and white. Travel west to the confluence of the Republican and Smoky Hill, where the Kaw begins, and the number drops to six. Head up either stream and, before long, only one species will be left—bur oak.

Named for the heavily fringed rim of the cap of its acorn, bur oak is a citizen of bottomland forests throughout much of the eastern United States. In the Upper Midwest, bur oak historically occupied a broad front along the forest–prairie continuum, occurring in stands of widely scattered trees called "oak openings." Along the Kaw and its tributaries, bur oak is mostly confined to edges of the ever-narrowing strip between sheltered stream valley and upland prairie.

Bur oak woodland can be found in the valleys of the Republican, Solomon, Saline, Smoky Hill, and their tributaries about as far west as the twenty-four-inch precipitation line. The oaks in these western stands occur in a belt along a terrace just above the cottonwood-populated floodplain. While relatively narrow, these stands are often packed with oaks, including some surprisingly large ones. The Kansas Biological Survey refers to these stands as Smoky Hills Oak Woodland.

Beyond this western front, which runs essentially from Holdrege, Nebraska, south to Hays, Kansas, bur oak is known from only a handful of pockets in some of the more remote nooks and crannies of the dissected loess plains of the Republican watershed. The most notable stand is in Bur Oak Canyon in southwestern Nebraska, a two-mile-long tributary of Driftwood Creek where three to four hundred trees are hunkered down below the windswept uplands. The isolation and concealment of this dense grove lends it an air of mystery and makes it a place of pilgrimage for a quirky band of enthusiasts we'll just call "oak people." I count myself among them.

Western bur oaks often take on a characteristic savanna physique, the trunks more thickset, the branching lower, and the crown more spreading. The leaves of these trees are dark green and shiny above with a whitish felt on the underside; when stirred by strong winds, they impart a flickering animation to the canopy. The ability of bur oaks to survive where other oaks wither is due in part to the thick, leathery texture of the leaves and bark dense enough to resist fire carried along by prairie grasses.

The lonelier the oak the more picturesque its frame, giving such trees an individuality that feels like personality. I can picture a number that have made an impression on me over the years. I suppose it is like a marine biologist studying whales. Spend enough time among these trees and you begin to recognize individuals. Bur oaks are the humpbacks of the prairie, breaching, here and there, up from the bluestem sea.

While a particular bur oak will have its particular beauty, a gathering of them brings a special magic to the landscape. There is something about the interweaving of oak groves and prairie pasture that is deeply pleasing. You can experience it in the rolling Rose Creek country of south-central Nebraska, which to me feels a bit like West Virginia, and in the valley of Paradise Creek in the Smoky Hills of Kansas.

Like bristlecone pine at timberline, bur oaks on the edge of the prairie grow slowly and, if given enough time, can attain a ripe old age. In the University of Nebraska State Museum there is a cross section of a bur oak tree that was cut down in 1918. It had a trunk four feet across and had been growing along Rose Creek, a tributary of the Little Blue River, for 312 years. John C. Frémont came across even bigger and likely more ancient ones in 1843. Traveling somewhere along the Solomon River, he encountered bur oaks with trunks "five and six feet in diameter."

Wherever oaks occur, and there are some four hundred species in the worldwide reach of the genus *Quercus*, they stir the heart. Across cultures and across millennia, oaks have been viewed with a mix of awe and affection and are celebrated as paragons of the best of human attributes—fortitude, fidelity, unhurriedness, and the like.

Such esteem extends to the bur oak, especially out on the plains, where it is the sole representative of its noble genus. You find it in the history of the Otoe people of Nebraska's Big Blue River country, who entrusted bur oak with their dead, bundling up the bodies of their loved ones and binding them to outstretched limbs. And you see it in the eyes of a plainsman or plainswoman when you ask about the stand of bur oak on their place.

I recently made the acquaintance of Tom Maupin, whose family's land in Russell County, Kansas, includes a bur oak–sheltering stretch of Paradise Creek. On December 15, 2021, the Maupin ranch took a direct hit from an extraordinary outbreak of wildfires fueled by drought-crisped pastures and what meteorologists call a derecho—a ferocious storm system that generated hundred-mile-per-hour winds. Thankfully no human lives were lost, although Tom and his wife Deb's home was burned to the foundation and nearly three hundred head of their cattle died or had to be put down.

But as Tom toured me around in his pickup, what he really wanted to talk about were his beloved bur oaks, some of which were killed or damaged by the wildfire. Amid all the loss this seventy-year-old man suffered, his question to me was, "Why would God let that fire tear up my trees?"

Cottonwood, specifically the eastern cottonwood, is the only tree species that occurs throughout the watershed of the Kaw, from its confluence with the Missouri to its headwaters on the High Plains. Cottonwoods are common members of muggy floodplain forests along the eastern reaches of the Kaw, but out on the far western tributaries they stand alone and shoulder more ecological and cultural significance than any other tree on the plains—more, perhaps, than any other tree in North America.

A cottonwood carries itself differently than a bur oak. Oaks have a certain stateliness about them even as they age, filling up their allotted air space with a dignified if burly symmetry. Mature cottonwoods can be a bit disheveled, with broken limbs and gaps in the crown and trunks missing big chunks of bark. I've seen many glorious cottonwoods, but some remind you of running into that old friend who has lived a very hard life or made some very bad choices.

Looks aren't everything, of course, especially when you're in need of shade from the sun, shelter from the wind, fuel for fire, and the prospect of water. Out in the deep plains, cottonwoods are a fountain of life.

But these trees also tug at the heartstrings. For travelers crossing the plains from the forested East, cottonwoods were the last familiar creature in a very foreign landscape. Edwin James of the Long Expedition of 1820 gave a wistful description of a grove scattered along the South Platte River in what is today eastern Colorado. The stocky stature of the cottonwoods and their wide spacing "revived strongly in our minds the appearance and gratifications resulting from an apple orchard, for which from a little distance they might readily be mistaken, if seen in a cultivated region." Other prairie travelers noted the refreshment of birdsong ringing out from cottonwood groves.

And then there are the leaves.

A cottonwood leaf is triangular in shape and dark green and lustrous on the upper surface. Like aspen in the Rockies, cottonwood leaves take on a goldenrod color at the end of summer, and a meandering plains stream lined with cottonwoods is a ribbon of bright yellow in the fall.

Also like aspen, cottonwood leaves "quake" with the slightest breeze and seem to be in perpetual motion. The quaking action is due to the shape of the petiole, the stalk that attaches the leaf blade to the twig. The petiole of a cottonwood leaf is two to three inches long and, instead of being round in cross section, is somewhat flattened on the sides. This slight modification is just enough to catch the wind and cause the leaf to shudder.

A lot of poetic language has been mustered over the years to describe the sight and sound of cottonwood leaves in the wind. Donald Culross Peattie wrote of cottonwoods casting "a pool of restless shade." "Dancing" is probably overused, but it is hard to come up with a more fitting description of the joyful wobbling of a cottonwood leaf, particularly when low-angling morning sunlight glints off the blade.

Compared to bur oaks, cottonwoods are fast growers and can put on heft and height in a hurry. At this writing, the reigning national champion eastern cottonwood, the largest in the United States, is growing near Beatrice, Nebraska; it stands 88 feet tall with a 108-foot canopy spread and a total trunk circumference of just over 36 feet. When I asked a forester for directions to this arboricultural celebrity, he felt compelled to add, "She's big and relatively old, but she's not too pretty."

When officially recognized as national champion in 2013, the Beatrice-area tree took the title away from a cottonwood planted in 1908 on a ranch near Studley, Kansas, which had assumed the honor in 2007, when a storm took down the previous national champion growing near Seward, Nebraska. An earlier title holder grew near Arapahoe, Nebraska, and measured ninety-six feet tall until it was felled in a windstorm. In case you haven't noticed, all these monarchs hailed from the country of the Kaw, a glory worth crowing about given the expansive native range of this tree.

Yet cottonwoods, like dynasties, can come to a sudden end. The floodplain habitat in which these trees flourish is dynamic and constantly changing due to periodic floods. This is a good thing when routine high water events rearrange the riverbed, creating sandbars and scouring bare other places where cottonwood seed can germinate and seedlings become established. But really strong floods can undercut the root systems of existing cottonwoods and take down even the largest trees. Unstable floodplain habitat, coupled with the relatively weak wood of cottonwoods, makes passage from big shot to driftwood an ever-present danger.

None of this seems to trouble a cottonwood tree, whose strategy is to grow quick, grow big, and throw tons of seed to the wind. As anyone who has spent time around cottonwoods can attest, these trees have the capacity by virtue of fluff-tipped, airborne seed to blanket the countryside with embryonic cottonwoods. The timing of this massive seed dump neatly coincides with the end of flood season, when freshly scoured habitat is most readily available.

This scheme has worked for millennia in the Great Plains but is now thwarted by nonnative species like Russian olive and Siberian elm, which have invaded floodplain habitat and shade out the exposed places needed to grow cottonwood seedlings. Flood control projects have also reduced the frequency and intensity of the high-volume surges that periodically open up new seed beds, further frustrating cottonwood regeneration. These challenges, coupled with water table drawdown by agricultural irrigation, are threatening the integrity and future of these unique and important woodlands.

Reports of cottonwood communities prior to Euro-American settlement describe open stands of trees scattered along waterways but also note occasional areas with extensive floodplain concentrations. These places were known as "Big Timbers." Two such areas were well-known in the western reaches of the Kaw. The Big Timbers of the Republican occupied a fourteen-mile stretch of bottomland immediately above what is today Swanson Lake reservoir in southwestern Nebraska. The Big Timbers of the Smoky Hill was located on the headwaters of that stream near the Colorado-Kansas border.

Groves of cottonwoods in full lustrous leafage stand out in dramatic contrast to the surrounding grassy plains, especially toward the end of a dry summer when the prairie turns the color of straw. The Big Timbers of the Smoky Hill must have been an especially impressive sight. The Pawnee name for the Smoky Hill River translates to "Big Black Forest" and the Anglo name for a stage station established in the vicinity was "Blue Mound," an allusion to the mirage-like appearance of the grove when viewed from a distance.

A vivid description of one of these groves comes from Eugene F. Ware, who was traveling along the Republican in the winter of 1865. He wrote, "These big timbers were all cottonwood trees averaging and exceeding two feet in diameter and located on average about one to every fifty square yards, without a particle of underbrush, but a dense growth of high bottom-grass." The open, grassy nature of these pristine cottonwood woodlands is echoed in the accounts of other early observers. Where such communities persist today, as along the Arikaree River in Colorado, the meadow-like understory is dominated by Indian grass and a few other tall grasses characteristic of eastern prairies.

It was not botanical curiosity that had Ware poking around the Big Timbers on January 20. He was part of a military campaign to "drive the Indians out of the Republican country." Such cottonwood groves were the favored camping grounds, especially in winter, of a number of plains tribes considered hostile by the US Army. Ware wrote that "Indian signs were everywhere" and concluded that "very great numbers of them had been hibernating through the heavy timbers scattered along the river at this place."

Big timbers and lesser stands of cottonwoods were crucial to the lifeway of plains tribes, a fact well-known to Gen. George Armstrong Custer and other military strategists. Beyond the obvious provision of shade, shelter, and firewood, cottonwood trees were a critical source of winter fodder for horses. Unlike bison, whose narrow, pointed hooves enable them to paw through snow to reach grass, horses, which have flat hooves, find the task much more difficult and sometimes impossible, necessitating an alternative or supplement to grazing in winter. At some point after the reintroduction of the horse to the Great Plains it was discovered that tender twigs and fresh inner bark stripped from the upper branches of cottonwood trees would sustain them in winter. The standard practice was to cut trees down entirely and harvest the twigs and bark for feed. It was the cottonwood that made it possible for horse nomads to occupy the central and northern plains in winter, and for village tribes like the Pawnee to make their annual winter bison hunt.

In the view of historian Elliott West, the horse was both gift and curse to plains tribes like the Cheyenne. While it gave them unfettered freedom throughout much of the year, it tethered them to cottonwood groves in the winter, which not only made them vulnerable to cavalry charges but also drew them into a perilous ecological bottleneck. As Indian horse herds grew, so did damage to cottonwood stands on the Republican and Smoky Hill, later exacerbated by incursions of wood-hungry Euro-Americans.

An economy based on bison and grass, horse and cottonwood, while elegant in simplicity, proved to be profoundly fragile.

Keeping grassland in good ecological condition often requires combat with woody plants. The chief adversary in our region is eastern red cedar, a very tough evergreen tree that can invade a wide-open prairie in such numbers as to turn it to woodland in just a few decades.

Unlike most invasives, eastern red cedar is not an exotic species; it is a native plant that became a weed through human agency. A cedar tree produces either male or female flowers, and female trees generate heavy crops of berries. A variety of birds feed on the berries and later pass the seeds, scattering them far and wide. Prior to Euro-American settlement, frequent and intense fires kept most seedlings of this fire-intolerant tree from getting a foothold in highly combustible prairies. But control of wildfire coupled with extensive planting of eastern red cedar as a windbreak tree on farms and ranches has enabled it to spread through grasslands like a dark tide.

A longstanding tradition of prescribed burning helps keep cedar in check in many tallgrass pastures of the Flint Hills. In the mixed-grass prairies of the Loess Canyons region of south-central Nebraska, where cedar encroachment has been a critical conservation concern, ranchers are finally regaining ground by fighting with fire.

You wouldn't know it today, but eastern red cedar was once an uncommon tree in the country of the Kaw, mostly limited to scattered occurrences on bluffs along waterways, presumably because grass cover in these rocky places was too thin to carry unchecked prairie fires into the trees. Some of these stands were notable enough to inspire names like Cedar Bluffs, Cedar Creek, and Cedar Gorge.

The most remarkable cedar-place in the country of the Kaw is Cedar Point in eastern Colorado. Located north of the town of Limon, Cedar Point is an elevated, westward-pointing prow of tableland where the High Plains surface falls away in cliffs to the Colorado Piedmont below. The name derives from the stands of cedar trees growing along and below the rocky rim, primarily on north-facing slopes. But these are not eastern red cedar, they are Rocky Mountain juniper. Scattered ponderosa pine grows here as well. What are these mountain trees doing so far out on the plains?

Isolated stands of Rocky Mountain conifers, including other species of junipers and pines, occur in the western Great Plains in association with escarpments and topographic breaks. University of Kansas biogeographer Philip Wells speculated such "scarp woodlands" are relicts of formerly more extensive forest systems that, during cooler and moister Pleistocene times, stretched eastward from the Rocky Mountains. Drying climate at the end of the Pleistocene caused the retreat of these woodlands to the west, with scatterings of trees left behind, clinging to areas of steep rocky habitat.

The edge of a High Plains escarpment would seem an exceedingly hostile setting for tree growth, but the rocky habitat provides deliverance from a tree's most bitter rival—grass. With a densely fibrous system of fine roots, grasses are tough competitors for water and nutrients in deeper soils but are not as capable of finding and keeping footing in stony substrates. The root system of a juniper or pine has a simpler branching pattern that is more adept at penetrating into the fractures, crevices, and shallow soil accumulations of exposed bedrock, giving these trees a competitive edge over the grasses.

You can't go any farther west in the watershed of the Kaw than Cedar Point, about five hundred miles from Kansas City as the crow flies. The same crow would only need to wing another fifty miles to reach the foothills of the Rocky Mountains. Not only can you see the Rockies from Cedar Point but you can also hear and smell the mountains as the wind works its way through the cedars and pines.

To be a tree in a land ruled by grass earns you peculiar standing in ecology and in human affairs. In the predawn hours of November 29, 1864, several hundred Cheyenne and Arapaho people were asleep in their tipis along Sand Creek, a tributary of the Arkansas River in southeastern Colorado. In a short while they would be set upon by 675 cavalrymen under the command of vainglorious monster John Chivington. Hours later, after the rifles and howitzers fell silent, about 230 Indians were dead, 150 of them women, children ("nits make lice"), and elderly men. Those who managed to escape the massacre, one of the most despicable acts ever perpetrated by the US government, struggled north across the open plains. The objective of their flight? The Big Timbers of the Smoky Hill.

It has been said that the Earth groans under the weight of our sins. If so, those Smoky Hill cottonwoods, sheltering the bereaved of Sand Creek, were wailing too.

FIVE

Setting Sail on the Star-Grass Sea

> Now the Prairie life begins!
> *Susan Shelby Magoffin, on departing Independence, Missouri,*
> *for Santa Fe, June 11, 1846*

The watershed of the Blue River in Jackson County, Missouri, is separated from that of the Little Blue by high ground known locally as the Blue Ridge. Engulfed today by Kansas City's eastern suburbs, the divide once provided passage to a place of both ecological transition and emotional transport.

At the north end of the Blue Ridge is Independence, Missouri, for many years the main spot where nineteenth-century westbound adventurers, entrepreneurs, freighters, gold seekers, and emigrants would "jump off" riverboats steaming up the Missouri and gear up for the overland part of their journey along the Santa Fe, Oregon, or California trails. After departing Independence, travelers would follow the divide in a southwesterly direction, eventually work their way down into the valley of the Blue River, cross it at a rocky ford, and, after a couple of miles, reach the boundary between the United States of America and Indian Territory, what is today the Missouri-Kansas state line.

It was along the Blue Ridge that travelers got their first immersion in prairie, which cloaked the divide and made it a grassy peninsula above the wooded valleys of the Blue and Little Blue, earning the upland the lyrical name of "Blue Prairie." For many, like Susan Shelby Magoffin, it was an exhilarating experience that received special note in their diaries and journals. The same for Edwin Bryan, a Kentucky newspaperman who was overcome by "a wild and scarcely controllable ecstasy of admiration" as he took to the Blue Ridge about a month before Susan Magoffin.

As journeyers continued westward into what is today eastern Kansas, trees began to fade from sight entirely, like the coastline as you head out to sea. Maritime language abounds in the writings of travelers struggling to fathom country that looked more like ocean than earth. It was an otherworldly, liquid landscape of rippling "star-grass," Wes Jackson's inventive name for the sunlight-trapping fabric of the prairie.

The initial thrill of a prairie excursion no doubt faded once adventurers realized just how long they would be at sea. Heading west from the Chouteau Trading Post, a few miles up from Kaw Point, in the spring of 1842, legendary explorer John Charles Frémont wrote romantically of setting out on "the ocean of prairie" but added with a tone of trepidation, "which, we were told, stretched without interruption almost to the base of the Rocky Mountains." Ecologists would eventually discern three principal types of grassland in the mid-continent—tallgrass prairie in the east, shortgrass in the west, and mixed-grass in the middle. Collectively, these make up the Central Grassland of North America.

Similar to the forest types of eastern North America, prairie formations are distinguished by the species of grasses that dominate the "canopy." For tallgrass prairie, these are typically big bluestem, Indian grass, and switchgrass, sod-forming species that reach four to six feet or more in height depending on the site and environmental conditions, with big bluestem sometimes topping out at ten feet. At the other end of the spectrum, shortgrass prairie is dominated by blue grama and buffalograss, low-growing species that produce vegetation six to eighteen inches in height. Mixed-grass prairie is a mixture of tall, short, and mid-height grasses; little bluestem, a bunchgrass averaging three to five feet in height, is usually the principal dominant. The exact composition and relative dominance of grass species within these formations is sometimes reshuffled locally depending on soil type, topographic position, and other environmental factors.

The east-to-west range of the three main prairie types is primarily a function of climate and is governed to a large extent by the seasonal dynamics of two ocean-spawned air masses. Tallgrass prairie occurs in the relatively rainy and humid eastern reaches of the country of the Kaw in a zone where weather systems born in the Gulf of Mexico drop twenty-eight inches or more of precipitation annually. Shortgrass prairie runs on moisture carried eastward from the Pacific Ocean, which, by the time the Rockies are done with it, provides less than nineteen inches per year for the semiarid western Great Plains. Mixed-grass prairie occurs in a middle zone, far enough out from the rain shadow of the Rockies to pick up some tropical Gulf moisture and receive twenty to twenty-four inches. The twenty-five-to-twenty-eight-inch precipitation zone is a region of transition between tallgrass and mixed-grass prairie.

There is growing recognition that our grasslands are also "wind-bounded"—that differences in prevailing wind speed from west to east shape the differences in species makeup among short, mixed, and tallgrass prairie. And, at the eastern edge of the mid-continental grasslands, strong westerlies historically influenced the boundary between prairie and woodland by affecting the intensity and movement of tree-consuming wildfire.

Like all biological "boundaries," the margins of these grassland systems are fluid, with significant back-and-forth shifts occurring in response to prolonged climatic patterns, as documented by the studies of Nebraska ecologist John Ernest Weaver during and after the great drought of the 1930s.

Given the seafaring metaphors used early on to describe our mid-continental grasslands, it is ironic that these ecological systems, far removed from the oceans surrounding North America, are still shaped by them—that the humid air soaking the shirt of a shrimper on his boat in the Gulf of Mexico will eventually flow north into the country of the Kaw and determine whether little bluestem or blue grama owns the uplands out along Sappa Creek.

Departing Independence for Santa Fe on the first day of November 1853, adventurer W. W. H. Davis embarked "upon the great prairie sea that stretches throughout the central region of North America." He would later write, "To a person who has never been upon the great American prairies, a trip across them can not be otherwise than interesting," adding, "upon these great plains a person experiences different feelings than when confined within cities and forest." He went on to declare, "A man who is not insensible to such influences can not fail to be made better and wiser by a trip across the prairies."

Let's set sail. . . .

The Flint Hills won't leave me alone. I've tasted the sweetness of this tallgrass landscape so many times and in so many ways that it haunts me like the hounded soul in the gospel song *Can't Shake Jesus*.

I first laid eyes on the Flint Hills as a boy, heading out from Kansas City on a family vacation to the Colorado Rockies. To a woodlander, the landscape was unsettling but, at the same time, alluring in the early morning light. Some two decades later I would botanize the Flint Hills from our home in Hesston, Kansas, and discover vistas so sublime I had to come back with Lynn and the kids (a cherished family photo shows Kendra, Karen, and Greg standing on a grassy ridge, the girls in *Little House on the Prairie Dresses* sown by Lynn, with only prairie and sky in the distance beyond). And now here in Lincoln, Nebraska, a hundred-plus miles to the north, the Flint Hills come after me each April in smoke from prescribed prairie burns, one time so thick my grandkids' school wouldn't let them outside for recess.

It is estimated that tallgrass prairie once covered 170 million acres of central North America and that this magnificent grassland has been whittled down to less than 10 percent of its former extent, making it one of the most diminished and imperiled ecological systems on Earth. Most of the original tallgrass country is now corn and soybean country.

The largest portion of the tiny portion of America's remaining tallgrass prairie is sheltered in the Flint Hills, where rocky soils and rugged terrain have mostly kept out the plow. The Flint Hills contains the largest contiguous tract of tallgrass prairie left on the continent, something on the order of 4.5 million acres. The Kaw and its tributaries drain most of the northern third of the Flint Hills. Rocky soils have also helped preserve significant amounts of tallgrass prairie in eastern parts of the Smoky Hills, where sandstone of the Dakota Formation is at or near the surface.

The majority of land in the Flint Hills is in private ownership, some owners multigenerational on the land, some absentee. The largest publicly accessible area in the Flint Hills is located south of the Kaw watershed in the 10,894-acre Tallgrass Prairie National Preserve, managed by the National Park Service in partnership with The Nature Conservancy.

The famous Konza Prairie Research Natural Area, owned jointly by TNC and Kansas State University, is situated within the Kaw watershed just a few miles south of the KSU campus. Encompassing 8,616 acres, the Konza is the largest tract of unplowed tallgrass prairie in North America dedicated to research and has been the setting for ecological studies by students and researchers from around the world.

Tallgrass prairie has not fared well outside of the Flint Hills and Smoky Hills, where soils are deeper and more amenable to cultivation. In the Glacial Hills and Osage Cuestas regions of eastern Kansas, tallgrass prairie can only be experienced in preserves like the Prairie Center near Olathe, Kill Creek Prairie in western Johnson County, and Rockefeller Prairie near Lawrence, each less than twenty acres in extent. In the watershed of the Big and Little Blue Rivers in southeastern Nebraska, tallgrass persists in only a handful of remnants, a number of which are protected through conservation easements held by the Wachiska Audubon Society of Nebraska.

Big bluestem, Indian grass, and switchgrass are the standard big three of midwestern tallgrass prairie, known historically as True Prairie or, simply, the Prairie. Subtle variations of this formula are found within the country of the Kaw. Prairie dropseed supplants switchgrass in the deep-soil Glacial Hills region of northeastern Kansas. Little bluestem bumps out switch in rocky upland areas of the Flint Hills. Historian James Malin referred to the Flint Hills as the "Bluestem Pastures," a poetic nod to the prevalence of big and little blue and to the importance of cattle grazing to the economy of this region.

In the Smoky Hills, soils underlain by Dakota sandstone support a big bluestem–switchgrass–little bluestem association recognized in Kansas as Dakota Hills tallgrass prairie and in Nebraska as Dakota Sandstone tallgrass prairie.

Within the greater tallgrass prairie region, lowland or wet prairie occurs in floodplain habitat where the soil is nearly always saturated or is at least temporarily inundated by flooding. In this densely vegetated community, prairie cordgrass dominates areas where soils are wet or poorly aerated, and big bluestem and Indian grass occur at the margins.

Stringers of tallgrass prairie occur as far west as the High Plains in sub-irrigated valley bottoms and terraces along larger streams and rivers, sometimes in association with open stands of cottonwoods. Big bluestem and Indian grass are the dominant species in these communities, which feel luxurious compared to the shortgrass prairie of the surrounding uplands.

Tallgrass prairie endures as substance and shadow in the country of the Kaw. Thankfully, horizon-to-horizon sweeps of this nearly lost American landscape can still be experienced in the Flint Hills and Smoky Hills. But even where it has been plowed up, its influence is still evident.

There is, of course, the economic muscle of prairie soils that now crank out corn and soybeans by the metric ton. Other traces are subtler. In *Kansas City and How It Grew, 1822–2011*, University of Kansas geographer James Shortridge describes the natural attributes that led to the establishment and eventual flourishing of this trading post at the merging of the Kaw with the Missouri. Chief among them, but underappreciated in his opinion, were the tracts of tallgrass prairie on what is today the south side of town, especially in the Waldo district, which provided pasturage for the thousands of oxen, mules, and horses needed by the overland trade of the 1800s.

KC may be famous for barbeque and jazz, but big bluestem helped put it on the map.

It seems no one can write about grasslands without quoting Willa Cather. Her word pictures of the prairie are not only beautifully crafted but also wonderfully accurate, not in scientific precision, but in look and feel. You find yourself rereading particular passages, amazed, not so much at what she said but at what she saw.

Yet Willa is almost always quoted out of context, at least ecological context. Her words are most often invoked to describe tallgrass prairie before the plow, although I have even seen her quoted in a preface to an article on Florida grasslands. The reality is, Cather's beloved prairie passages portray a very distinctive type of grassland—the mixed-grass prairie.

It took a while for ecologists to actually see the mixed-grass prairie. For many years the grasslands of central North America were thought to consist of a "prairie formation" on the east and a "plains formation" on the west. The prairie formation denoted the tallgrass prairie of the Midwest, while the plains formation encompassed the grasslands between the tallgrass prairie and the Rocky Mountains. Prevailing opinion was that the one passed into the other through a broad transition zone.

Pioneering ecologist Fredrick Clements, classmate of Willa Cather at the University of Nebraska, was the first to recognize the "mixed prairie" as a discrete and stable plant community; he described it in a 1920 publication as "composed of the dominants of both prairies and plains." Clements's concept is now universally accepted by ecologists and usually rendered "mixed-grass" or "mid-grass" prairie.

Broadly defined, mixed-grass prairie ranges from the Canadian Prairie Provinces south into central Texas. Ecologists eventually recognized a northern and southern division of the mixed-grass prairie; the Colorado-Wyoming border constitutes the approximate boundary between the two. The northern portion is dominated by cool season grasses that undergo peak growth during the spring, while warm season grasses prevail in the southern portion.

Cather's famous prairie novels *O Pioneers!* and *My Ántonia* are set in the fictional Nebraska villages of Hanover and Black Hawk, but her descriptions of the land are drawn from her years in Red Cloud, where she lived from 1883, when she arrived as a nine-year-old, until she left for college in Lincoln. Situated in the Republican River valley, Red Cloud is located at the northern end of the southern mixed-grass prairie, which extends from south-central Nebraska through the middle parts of Kansas and Oklahoma and onto the Edwards Plateau of Texas.

While she never mentions it by name, Cather perceptively sensed the influence of one particular species of grass in the prairies around Red Cloud, a grass that defines the southern mixed-grass prairie as the saguaro does the Sonoran Desert—little bluestem: "I felt motion in the landscape; in the fresh easy blowing wind, and in the earth itself, as if the shaggy grass were a sort of loose hide, and underneath it herds of wild buffalo were galloping, galloping..."

Unlike the sod-forming big bluestem that reigns in tallgrass country, little bluestem is a mid-height bunchgrass and grows in individual clumps, giving the mixed-grass prairie a look for which "shaggy" is remarkably apt. Cather's sharp eye even saw the distinctive way little bluestem catches the wind—at a choppy buffalo gallop rather than the fluid stallion stride of big bluestem. As fall approaches, little bluestem impresses more of its character upon the landscape, acquiring a tone of red Cather described as "the color of wine stains": "All those fall afternoons were the same, but I never got used to them. As far as I could see, the miles of copper-red grasses were drenched in sunlight that was fiercer than at any time of day. The whole prairie was like a bush that burned with fire and was not consumed."

Accompanying little bluestem in the mixed-grass prairie are other grasses of similar or smaller stature. The graceful sideoats grama, with seed clusters that resemble Indian lances arrayed with a row of pendant feathers, is often second in abundance. On drier sites, blue grama, hairy grama, and buffalograss—species of the shortgrass prairie—gain importance. Attending these grasses are a wealth of wildflowers, many shared with tallgrass and shortgrass prairie but others essentially limited to the mixed-grass prairie.

The largest and finest tracts of this distinctive grassland are preserved in the rocky uplands of the Smoky Hills, especially the western two-thirds, where limestones of the Greenhorn and Fort Hays rock systems dominate the surface geology. Extensive tracts also flourish on uplands in the Loess Canyons region of south-central Nebraska. The Willa Cather Memorial Prairie south of Red Cloud provides a beautiful rolling expanse of 612 acres of mixed-grass prairie with almost two miles of hiking trails.

Northern mixed-grass prairie makes a limited appearance within the watershed of the Kaw, occurring in deep loess soils and on rocky uplands in the High Plains region of southwest Nebraska, northwest Kansas, and an adjacent bit of Colorado. This particular expression is characterized by threadleaf sedge, a densely tufted, low-growing grasslike plant, in association with blue grama of the shortgrass prairie and needle-and-thread, a mid-height, cool-season grass.

To my eye this is one of the most handsome grassland associations in all of the Great Plains, especially in spring. It cloaks rocky hills bordering the valley of Frenchman Creek as it works its way down from the tablelands to meet the Republican at Culbertson, Nebraska. It also occurs in the scenic Arikaree Breaks region in the extreme northwest corner of Kansas.

But it is little blue that owns the middle country of the Kaw. You can run into this distinctive grass in many places across the United States. I've seen it in serpentine barrens in Maryland, cedar glades in the Ozarks, and in the foothills of the Rockies. "But the true home of little bluestem," according to ecologist David Costello, is the mixed-grass prairie country of the central Great Plains.

We will never know how many millions upon millions of bison were roaming the plains when awestruck Euro-American explorers witnessed the immense herds of yesteryear. Appearing in swarms so vast and so dense as to completely darken the land, their numbers defied calculation and description.

Ironically, these stupendous throngs of massive herbivores flourished on the sparest of herbage—the shortgrass prairie.

The term *shortgrass* has been used broadly to refer to the westernmost grasslands of the Great Plains, from Canada south into Texas, but ecologists now use the term more narrowly to distinguish an ecological system dominated by blue grama and buffalograss. Thus defined, the general range of shortgrass prairie includes most of Colorado and New Mexico east of the Rockies plus western Kansas, the Oklahoma Panhandle, and the western part of the Texas Panhandle. The Republican and Smoky Hill Rivers drain the northern sector of this ecological system.

If there is beauty in simplicity, then shortgrass prairie is the most beautiful of grasslands. While a local expression of tallgrass or mixed-grass prairie will be dominated by a particular species of grass, there will be several other characteristic grasses that join it to form the fabric of the vegetation. Shortgrass prairie is essentially a two-grass grassland. Local soil conditions and grazing history can tip the balance back and forth between blue grama and buffalograss, but no other grass comes close to breaking into the inner circle.

Shortgrass prairie also has the thinnest and least diverse wildflower component of the three main prairie types. As a result, an expanse of shortgrass prairie looks for all the world like a gigantic lawn, especially when greened up by favorable rains. Traversing the upper Republican watershed in June 1843, Frémont noted the prevalence of buffalograss on the uplands, "giving to the prairies a smooth and mossy appearance."

Elegant is not a word that springs to mind when contemplating a bison, but there was a profound ecological elegance in the relationship between bison and shortgrass prairie. I wonder if there is another place on Earth where the grazer and the grazed were so perfectly fitted to one another.

On the bison side of the ledger, the foliage of the two grasses is not only highly nutritious but also holds its nutritional value well past the end of the growing season, especially blue grama, providing sweeping stands of cured hay for fuel during the winter. On the grass side, grazing can actually spur plant growth.

The region of a plant from which new growth is generated is called the apical meristem. In blue grama and buffalograss, the apical meristem is located at or near the soil surface and is protected by enveloping leaf sheaths. This enables the plants not only to withstand close cropping but to actually benefit from it. The removal of old leaves by grazing stimulates new growth, the tufted blue grama by arching tillers and the carpet-forming buffalograss by creeping stolons.

Additionally, studies have shown that when cattle graze blue grama and buffalograss, they consume seeds along with foliage. Seeds not crushed by chewing pass through the gut and out the back end of the beast, gaining transport to new territory, partial digestion of germination-inhibiting substances, and a manure-fueled boost to initial growth. The Great Plains cattle industry of today is founded upon a mutualism hammered out in millennia past between bison and this duo of grasses.

The term *steppe* is sometimes used for the semiarid grasslands of North America, including shortgrass prairie. This evocative Russian word was first applied to the grasslands of Eastern Europe and central Asia, and is the favored label of those who think about grasslands on a global scale. The word *prairie* originated with French fur traders who needed a name for the astonishing land of grass they encountered in central North America. Having never heard of the steppes of Russia, the trappers turned to the Old French *praerie*, derived from the Latin *pratum*, meaning meadow or pastureland. I'm going with *shortgrass prairie* since it is more familiar to American ears, but *shortgrass steppe* has more music to it.

While tallgrass, mixed-grass, and shortgrass are the chief prairie types in the country of the Kaw, dune fields in the western reaches of the Republican River support an additional grassland formation of note—the sandsage prairie.

Like Nebraska's celebrated beef and cabbage Runza® sandwich, sandsage prairie is an acquired taste. If you took a poll, and excluded birders looking to add Cassin's sparrow to their life list, sandsage prairie would rank as one of the most poorly known and least appreciated ecological systems in all of the Great Plains. But I have come to admire sandsage prairie for its specialized flora, high-summer exuberance, and capacity to heal from the wounds of the wind.

The sandhills of the Republican River share many grass and wildflower species with the Nebraska Sandhills to the north. The primary difference is the prevalence of sand sagebrush or sandsage, which occurs in sporadic patches within the main body of the Sandhills and in larger belts along the southwestern edge, but never in the extensive tracts found in the Republican Basin.

Sandsage is a much-branched shrub that grows up to three feet tall and two feet across. The gray-green foliage is aromatic and releases a pleasant odor when rubbed or crushed. A diagram in Weaver and Albertson's *Grasslands of the Great Plains* shows the root system of a ten-year-old shrub with a dense, well-branched network of roots in the upper three feet of soil and a sparingly branched taproot that extends eight feet below the surface. The layout not only anchors the plant but also captures precious moisture before it quickly percolates down through the sandy soil.

Sandsage is one of the few shrubby sagebrushes (genus *Artemisia*) with the ability to resprout after fire, a valuable attribute when growing in the midst of combustible grasses. Although fire can kill big sagebrush, the shrub that dominates so much of the West, it seems to actually stimulate sandsage, and scorched plants resprout "profusely."

Sandsage prairie has a basic structure of evenly spaced sandsage shrubs with the intervening ground dominated by grasses. The grass component typically has a canopy of taller grasses (three to six feet) intermixed with mid-height grasses (eighteen inches to three feet) with an underlayer of short grasses and sedges. This vegetation is properly classified as shrub-steppe rather than grassland, but the term *sandsage prairie* has wide and longstanding use and reflects the codominance of prairie grasses with sandsage. Many ranchers simply call both the plant and the plant community *sandsage*.

On relatively level to gently rolling terrain, prairie sandreed and needle-and-thread are the dominant grasses that occur with sandsage; sand bluestem and sand dropseed are also important. Prairie sandreed and sand bluestem, the tallest of these, reach five to six feet at their prime in late summer, towering above the rest of the vegetation. The low-growing blue grama and sun sedge usually make up the underlayer.

Things get more interesting where the topography is stronger and the dunes choppier. With sharper crests and steeper slopes, these dunes are more vulnerable to wind erosion and often have areas where the vegetation has been thinned and bare sand is exposed. In severe cases, the wind scoops out a crater-like depression called a blowout. Blowing sand, high light intensity, and high temperatures make the interior of a blowout a very hostile environment for plant life.

There is a constant give-and-take between wind and vegetation in the sandhills, facilitated by a succession of plant species, each adapted to a particular stage in the progression from blowout to stable vegetation. The first to invade the bare, loose sands are a few diminutive annuals, followed by blowout grass and lemon scurfpea, perennials that spread by rhizomes—horizontal stems that take off laterally below the surface of the sand and send up new shoots along the way, sometimes twenty to forty feet away.

As these early pioneers spread throughout a blowout, they begin to take the edge off the harsh environment and usher in the next stage of recovery, led by sandhills muhly. This tufted, wiry grass also spreads by rhizomes and forms matted clumps about ten inches tall. The central part of the mat often dies out in older plants, which leaves an open crescent that is perfect for catching blowing sand and holding it in place; this further stabilizes the blowout habitat and prepares it for the return of sandsage. This middle stage is easily detected in late summer when the flower clusters of sandhill muhly take on a purplish-red color and impart a striking crimson tone to choppier parts of the landscape, visual assurance that recovery is under way.

In *Saving the Prairies*, historian Ronald Tobey describes the seminal role of the "grassland school" at the University of Nebraska, which produced a network of scientists whose research in the prairies of the mid-continent would lay the groundwork for the emerging science of ecology. Dating back to 1895, these luminaries include Charles Bessey, Roscoe Pound, Fredrick Clements, John Weaver, Homer Shantz, Fredrick Albertson, and a host of their students and coworkers.

Many of the study sites of these pioneering ecologists were in the country of the Kaw, from the tallgrass prairies of the Big Blue River region to the mixed-grass of the Smoky Hills to the shortgrass and sandsage of the Colorado High Plains. The patterns and principles they deciphered in the realm of the Kaw provided foundational insights into the marvelous workings of North America's star-grass sea.

Though not the only river to flow across the prairies of the Great Plains—there are the Yellowstone, Niobrara, Platte, Arkansas, Cimarron, and Canadian—the Kaw is the only one of these to arise on the plains and not in the mountains. Born in shortgrass and sandsage, gaining heart in the mixed-grass, mellowing out in the tallgrass, the Kaw is truly America's prairie river.

SIX

A Kaw Florilegium

> I speak often of the grass and the flowers ...
> *John Charles Frémont,* Memoirs of My Life

In August 1768, HMS *Endeavour* set sail from England on a voyage of discovery around the world. Famously commanded by Captain James Cook, the expedition's company included Joseph Banks, a wealthy young man with a passion for botany. Three years later, Banks was back home with over thirty thousand plant specimens in his possession, more than thirteen hundred representing species new to science. He soon commissioned a team of artists to create illustrations of his discoveries. The work, not completed in his lifetime, would eventually depict 743 species and become known as *Banks' Florilegium*—a gathering of his flowers.

The sum total of plants that grow naturally in a given region is called its *flora*. The flora of the country of the Kaw includes trees and grasses, which give structure and definition to the main types of vegetation, plus a host of other plants that are neither trees nor grasses. There are woody shrubs and vines, grasslike plants (graminoids) such as sedges and rushes, ferns and fern allies (pteridophytes), and mosses and liverworts (bryophytes). But the real diversity comes from what woodland ecologists call herbs and prairie ecologists call forbs. Most everyone else calls them wildflowers.

A complete accounting of the wildflowers of the county of the Kaw would need to cover hundreds of species. Like the artists working for Sir Joseph Banks, we'll have to keep it to a florilegium.

It should not be surprising that a literary world of faeries and hobbits and motorcar-crazed toads springs from forested places. The imagination seems most deeply stirred among the trees. It is in the woods especially that we find our way to *Elfland*, "that place," wrote G. K. Chesterton, "with a most dark and deep and mysterious levity."

The most enchanting season to be afoot in the great deciduous forests of eastern North America is early spring, when the first of the woodland wildflowers begin to bloom. These plants, at their western limits in the country of the Kaw, bear olden, affectionate names like spring beauty, bloodroot, Dutchman's breeches, wake-robin, toothwort, fawn-lily, Mayapple, Jack-in-the-pulpit, Solomon's seal.

There is much about these plants that pierces the soul. First, there is the charm factor. Of all the wildflowers that grace the eastern United States, none are as winsome as these woodlanders. Most are what you might call dainty—diminutive in stature, with delicate flowers and foliage. Some, like Jack-in-the-pulpit and Mayapple, are more robust but possess an endearing oddness, like a goofy uncle, or a troll. Each has its own unique personality and appeal.

Then there is their timing, cleverly blooming at the threshold of spring, when the human heart is at its most vulnerable. Delivering the year's first flowers, these plants have long been cherished as emblems of renewal. Walking the woods of his Kentucky farm, Wendell Berry takes comfort in "Mayapples rising out of old time" and "the resurrection of bloodroot from the dark." They are also a trigger of nostalgia and longing. For my wife, Lynn, Dutchman's breeches stir memories of her beloved granny Hazel, who would tear up at the sight of them in the awakening woods around her Hannibal, Missouri, home.

The flowers of most spring woodland herbs are white or whitish in color and have a shape like an open bowl. The color and shape likely serve to attract insects that act as pollinators: white stands out against the brown and green background of the forest floor, and the simple radial symmetry provides basking sites for insects, particularly bees and flies, that are on the wing early in the season and need a break from rough flying conditions.

One exception to this white-bowl flower pattern is timber phlox, also known as wild sweet William, a plant with lavender-colored flowers in which the petals are fused into a long tube that opens into a platform of five spreading lobes. Timber phlox occurs in woodlands throughout the eastern United States, and its far western limit is in the gallery forests of the Flint Hills. It is the most strongly scented of the early spring bloomers, which no doubt helps attract searching insects to its trumpet-shaped flowers. Studies of the pollination ecology of timber phlox in a Flint Hills woodland found that butterflies and especially moths, rather than bees and flies, made up about 90 percent of its floral visitors.

The bloom time of these woodlanders is as brief as it is early; most wrap up their display as the tree canopy fills in with the year's new leaves and the dappled shade deepens to full. The majority also go dormant and disappear completely belowground for the rest of the growing season. Their fleeting presence has led ecologists to refer to them as "spring ephemerals," although all of these plants are perennials and some are very long-lived. Regardless, it seems like you no sooner get reacquainted than they're gone.

"Timbered country has spring subtleties, rising sap and first bud, florets on trees and bushes, half-hidden flowers among the rocks and leaf mold," wrote Hal Borland in his 1956 memoir, *High, Wide, and Lonesome*. "Spring on the plains has little more subtlety than a thunderstorm."

Spring is a fickle season throughout the country of the Kaw, but it is especially volatile out on the western plains, where wild temperature swings are the norm and a May blizzard no big surprise. Still, the shortgrass prairie has its own guild of early spring wildflowers. Like their woodland counterparts, they need to shake off winter and get their business done before they are shaded out by taller plants. But instead of dealing with fifty-foot-tall bur oaks, they're racing against the new growth of blue grama and buffalograss, which in a good year might top out at twelve inches.

Easter daisy, sand lily, and Nuttall's violet are the vernal vanguard of the shortgrass flora. Small in stature and with a relatively brief flowering period, they can be easily missed among the curling, winter-worn grasses. But what they lack in size and seasonal longevity they make up for in floral output, brightening the dun-colored prairie like bouquets scattered by a rowdy wedding party.

These plants are well adapted to the challenges of the shortgrass spring. Blooming early enables them to take advantage of the higher levels of soil moisture provided by late snows and early rains. Keeping low keeps delicate flower parts out of the wind, which in spring can be fierce enough to make your ears ring and eyes water. Going dormant after flowering allows them to escape looming summer dryness and competition from more robust grasses and forbs.

Easter daisies wake up as early as the middle of March and offer the first hope that winter is losing its grip on the western plains. The plant grows as a compact, cushion-like mound, at most three inches tall. The flower buds, formed the previous fall, all open up within a few days; so, while the blooming season is rather short, the plant all but disappears under a decking of stemless, inch-wide daisies. Easter daisies are not a plant you go out to look for on purpose, like the year's first woodland wildflowers. More often than not you just stumble across one tucked up against a clump of blue grama, although an early spring outing will occasionally reward you with the sight of a small colony.

A patch of sand lilies looks like an outbreak of subterranean stars. The plant consists of a tight tuft of grassy leaves, four to eight inches long, out of which arises a spray of white, six-parted flowers. The lily-like flowers, as well as the leaves, emerge from a deep-seated rootstock, which places the ovary where the seeds are formed well below ground level. According to Colorado botanist Bill Weber, the flower buds of the following season push the seeds of the previous year's flower up to the surface, where they get scattered around by ants and other foraging insects.

With more than five hundred species worldwide, violets (genus *Viola*) turn up in all kinds of places. Nine species occur in the country of the Kaw. Seven sport the classic purple-blue flowers you expect of violets, but two have yellow flowers—downy yellow violet and Nuttall's violet. Exemplifying the ecological amplitude of the genus, downy yellow violet occurs in the Kaw's eastern woodlands while Nuttall's violet is a plant of the shortgrass prairie. Both are early spring bloomers. Writing of Nuttall's violet in his delightful book *Jewels of the Plains*, Claude Barr praised the "scintillating reflections of its textured gold petals."

In the spring of 1910, ten-year-old Hal Borland was exploring his family's new homestead on the plains of eastern Colorado. They had recently come to the shortgrass frontier from the comparatively mild environs of eastern Nebraska, and the boy, who would grow up to become a journalist and one of America's most prolific nature writers, was already reading the land. Poignant encounters with two of our springtime charmers are found in *High, Wide, and Lonesome*: "The hills turned greener every day, and in one little hollow not a hundred yards from the house I found a patch of yellow violets, tiny golden violets that I picked and carried to Mother. She said they were the prettiest flowers she'd seen since we left home, and she put them in a water glass and set them on the window sill."

Nuttall's violet wasn't the boy's only floral discovery that spring. Up the slope behind the house, he found "the grass twinkling with sand lilies." He picked a bunch of the flowers for his mom, but they wilted before he got to the door.

Anytime is a fine time for a ramble through a tallgrass prairie. But certain times are finer times than others.

In older scientific literature, ecologists would describe the "seasonal aspects" or "seasonal societies" of plant communities—natural groupings of wildflowers with similar peak flowering periods. In their landmark 1934 paper titled, underwhelmingly, "The Prairie," John Ernest Weaver and T. J. Fitzpatrick of the University of Nebraska described four distinct seasonal aspects in the tallgrass prairie—prevernal, vernal, estival, and autumnal. About fifty years later, University of Kansas ecologist Kelly Kindscher discerned eight fine-scale groupings in the tallgrass prairies around Lawrence, composed of plants that share a common life-form, flowering season, and ecology. He called these assemblages "guilds."

This imaginative term harkens back to medieval Europe, when artisans would associate together in societies to further their specialized trade or craft. The use of "guild" as an ecological concept goes back to the 1890s, but Kelly was the first to apply it in grassland ecology.

In researching his monumental work, *The Great Platte River Road*, historian Merrill Mattes consulted hundreds of diaries, letters, and other first-person accounts of emigrants who traversed the eastern third of the Oregon Trail from the 1840s into the 1860s. Setting out from Independence, Westport, or other "jumping-off" points, most began their journey in May or early June, when the grass had greened enough to nourish their draft animals.

That had them wading through the tallgrass prairies of the country of the Kaw during the heyday of Kelly's "spring forbs" guild, when the prairies were at their most flowery, which led to many an enraptured description of the landscape: the "Eden of America," wrote one; the "Garden Spot of the World," another.

A number of nineteenth-century botanical explorers also found themselves in the country of the Kaw at this heady time of year. Although anxious to reach unbotanized regions beyond the Great Plains, they, too, were struck by the spectacle. Departing Westport on May 22, 1843, Charles Andreas Geyer, "an assiduous German botanist" on his way to Oregon Territory, commented on "the beautifully undulating prairies of the lower Kanzas, stretching themselves, as if endless, along the horizon." Geyer added, "It is a charming sight, in the months of May and June . . . to behold these prairies teeming with flowers."

Another German, Friedrich Adolph Wislizenus, was in the vicinity of today's Olathe, Kansas, in May 1846 and observed, "The prairie over which we traveled looked more beautiful than I had ever seen it. The grass had all the freshness of spring, and the whole plain was so covered with flowers . . . that it resembled a vast carpet of green, interwoven with the most brilliant colors."

If you're a modern-day tourist from Germany or, for that matter, Los Angeles, and want to see the tallgrass prairie at its most exuberant, I suggest you aim for the late-spring-to-early-summer days of May and June. This is the spangled season, when a rich variety of colorful forbs, most around a foot tall, are on full display in the uplands because the dominant prairie grasses haven't yet put on much height. Not that the landscape isn't beautiful the rest of summer and in the fall, because it is, but with the grasses at full stretch and the yellows and golds of sunflowers, silphiums, and goldenrods swamping the rest of the floral spectrum, the prairie gets visually quieter as the growing season progresses.

You'll need a good wildflower guide. If you're new to plant identification, I recommend one where the entries are arranged by flower color. Here's what you'll likely see:

White—prairie larkspur, New Jersey tea, spider milkweed
Yellows (including cream and orange)—plains wild-indigo, prairie ragwort, prairie parsley, golden Alexanders, fringed puccoon, hoary puccoon

Reds (including pink and lavender)—prairie phlox, pink poppy mallow
Blues (including purple and violet)—bracted spiderwort, blue wild-indigo, ground plum

Of all the stars of this season, none shines brighter than prairie phlox. Both Geyer and Wislizenus specifically noted the beauty and abundance of this wildflower in the Kaw's tallgrass prairies. This is the open-country cousin of the timber phlox of eastern woodlands and has a similar trumpet-like flower with an elongate floral tube that flares out into a flat platform of five spreading lobes. The flower color of prairie phlox ranges from rose pink to near red to purple and, with a tendency to grow in spreading colonies, lights up the prairie with bright drifts of color in May and June. The vivid flowers, often highly fragrant, are sought out by at least twenty different species of butterflies.

Color, fragrance, and butterflies! No other wildflower brings such liveliness to the tallgrass prairie. The Reverend Charles Harrison of York, Nebraska, wrote of driving a wagon through Minnesota tallgrass prairie in the 1850s and witnessing "thousands of acres" of prairie phlox in bloom: "I can never forget the sight. All around I was greeted with those happy, smiling faces, and all the air was incense laden. Far as the eye could reach I was surrounded by those great masses of loveliness. I was in raptures."

University of Kansas botanist William Chase Stevens eulogized "the charm and mystical influence" of the tallgrass prairie in his 1948 book, *Kansas Wild Flowers*. His outburst, a bit emotional for an academic botanist, was incited by remembrance of a landscape "besprinkled" with prairie phlox. Professor Stevens had tasted spangled season in the tallgrass and fallen under its spell.

The stature of John Charles Frémont as an explorer of the American West becomes clear when you realize just how many places this side of the Mississippi bear his name. The list of "Fremont" toponyms includes four counties, a dozen towns and cities, and an assortment of peaks, passes, canyons, and rivers. But the colorful Frémont was also one of the most important botanical travelers of nineteenth-century America, as evidenced in the over twenty eponyms bestowed by scientists on plants for which Frémont collected the first specimen of what they would recognize as a new genus (*Fremontia*; *Fremontodendron*) or a new species (*fremontiana*; *fremontii*).

Frémont commanded five legendary exploring expeditions to the West during his career, each of which launched from the environs of today's Kansas City. You can learn a lot about the country of the Kaw before settlement by reading Frémont's reports on his first (1842) and second (1843–1844) expeditions; written like travel adventures, they contain wonderful descriptions of plants, wildlife, indigenous peoples, and the look and feel of the land. The official documents for the three other expeditions are mostly concerned with the country west of the Great Plains.

Although primarily tasked with making maps, Frémont paid close attention to the plants he encountered on his explorations. He often wrote about them in the narrative of his reports and collected specimens when he could for the study of grateful botanists associated with eastern institutions. Frémont sent plant specimens from his first expedition to Dr. John Torrey in New York, professor of chemistry at Princeton and Columbia and one of the foremost American botanists of the day. Torrey prepared a catalog of Frémont's plants that was appended to the expedition's official report, published in 1845. The list includes a number of tallgrass prairie wildflowers like prairie phlox, encountered on the "Big Blue river of the Kansas."

It was on Frémont's second expedition that the Smoky Hill River earned a notorious place in the history of American botanical exploration. Departing Westport in May 1843, the party followed the Kaw and then the Republican out to its headwaters in what is today northeastern Colorado. They continued on to the mountains and eventually reached the Pacific Coast in Oregon; then they looped south through California and headed back east through the Great Basin and across the Rockies.

The explorers hit the upper reaches of the Smoky Hill River in July 1844. On the night of the thirteenth, while they were in camp in today's Logan County, Kansas, a succession of powerful thunderstorms dumped heavy rain into the watershed. The deluge caused the Smoky Hill to rise suddenly and flood the encampment, ruining more than half of the plant specimens collected in the country between the Rockies and the Pacific, precious specimens eagerly awaited by Dr. Torrey and other botanists back East. It was a huge loss to the scientific community and was devastating to Frémont, who lamented in a letter to Torrey, "I have never had a severer trial of my fortitude."

But the Smoky Hill redeemed itself by offering up from its embrace two beautiful plants, an evening primrose and a clematis, each of which would prove to be new to science and would bear the name of the Great Pathfinder. Botanist Sereno Watson of Harvard published *Oenothera fremontii* in 1873 and *Clematis fremontii* in 1875 based on specimens collected on Frémont's second expedition, probably a day or two after the flood as they continued their trek back to "the little town of Kansas."

Both of these striking plants are specialties of the Smoky Hills and signature species of a guild of wildflowers that turn gravelly prairie ridges and hilltops into colorful pastures in May and June. These communities, associated with weathered exposures of Greenhorn or Fort Hays Limestone within the context of mixed-grass prairie, feature showy species like plains skullcap, bitterweed, narrow-leaf bluet, plains penstemon, Texas sandwort, purple coneflower, and milkwort. These are plants of modest stature, most less than a foot tall.

Watson considered Frémont's evening primrose a distinct species, but later botanists recognized it as a western variant of the Missouri evening primrose that graces rocky ridges in the Flint Hills and prairie-like glades in the otherwise forested Ozarks of Missouri and Arkansas. Frémont's namesake has the same soft yellow, saucer-like flowers as Missouri evening primrose, but they are only about a third the size. The overall stature of the plant is smaller as well, more condensed and tufted, and the leaves are very narrow and covered with silvery hairs, all adaptations to the drier environment of the Smoky Hills.

Frémont's clematis is one of the most distinctive wildflowers in the flora of the Kaw. Each plant is a dense cluster of stout stems, six to twelve inches tall, crowded with big oval leaves and tipped with nodding, bonnet-shaped flowers, purple gray in color and about an inch long. The charming flowers have a thick, leathery texture to them, as do the leaves, hence the common name "leather flower."

But the beauty of Frémont's clematis extends well beyond the blooming season. As the flowers fade they are replaced by attractive clusters of achenes—small, dry, one-seeded fruits, each tipped with a graceful plume. And, as summer progresses, the foliage dries to a pleasant saddle brown.

The leathery leaves persist well into winter, with tissue between the veins eventually weathering out and leaving behind a filigreed lacework.

Frémont's clematis is also something of a tumbleweed. Late in summer the main stem breaks away at ground level, and the dried plant, foliage and achene clusters intact, catches the wind and cartwheels across the uplands, scattering seed along the way. It is not uncommon to see large numbers of the dried plants piled up in draws and ravines as you travel through limestone areas of the Smoky Hills, evidence, even in winter, that this one-of-a-kind wildflower graces the neighborhood.

Although first discovered in the Smoky Hills, Frémont's clematis also occurs in a scattering of Ozark glades and, remarkably, isolated spots in Tennessee and northwestern Georgia. But the biggest, boldest stands of this wildflower are in the Smoky Hills, especially in Post Rock Country, where it is the floral emblem of this unique landscape and a tribute to one of America's most important botanical explorers.

The thing I find both refreshing and jarring about sandsage prairie is just how flashy it is. You would expect this plant community, which occupies dune fields of almost pure sand, to have a Sahara-like austerity to it. Even more surprising is that sandsage prairie is at its showiest during the height of summer heat and drought, when most other grasslands are in the doldrums.

In 1939, Francis Ramaley, a botanist at the University of Colorado, published a seminal paper on the ecology of sandhill vegetation in northeastern Colorado. In it he recognized a series of seasonal societies. The final one of the growing season was his "serotinal-autumnal" aspect, which commenced in late July and peaked in mid-August. "In most years," he wrote, this is "the time of greatest flower abundance."

The signature wildflower of this season is silky prairie-clover, a beautiful shrub-like plant about a foot tall with silvery foliage and a profusion of four-inch-long spikes packed with tiny rose-purple flowers. At its peak bloom in mid- to late July, silky prairie-clover is one of the most noticeable plants of the sandsage prairie, "abundant enough," Ramaley observed, "to produce a pinkish landscape." The sight made such an impression on him that he suggested the season could be called the "prairie-clover period."

Accompanying silky prairie-clover are a host of other summer-blooming wildflowers. A number are tall and lanky, two to three feet in height. These include showy gilia, with its clusters of long white trumpets, and bractless mentzelia, lifting up fragrant, afternoon-opening stars. Others, like sandhills fleabane and othake are shorter and occupy the understory. Then there is the spectacular bush morning glory, a sprawling plant of arching stems that form hemispherical mounds up to three feet tall and six feet in diameter, covered with pinkish-purple, funnel-shaped flowers two inches across when fully opened.

Along with these bold wildflowers, the dramatic sand bluestem and prairie sandreed are coming into their prime during the prairie clover period. Reaching five to six feet in height when in flower, these towering grasses look like they belong in the well-watered prairies of the Midwest and stand in sharp contrast to the ankle-high blue grama and buffalograss of the surrounding shortgrass prairie.

What gives? How can there be such botanical extravagance at the height of summer in a semiarid climate? It has to do with the unique soil–water relationships of sandy soils.

In the dune fields that support sandsage prairie, rainfall infiltrates rapidly and percolates deeply with little or no runoff. Evaporation quickly dries out the surface but only to a depth of an inch or two, leaving a layer of dry sand that reduces further evaporation.

Below this dry surface mulch, the sandy soil is usually well supplied with moisture that is more easily taken up by thirsty plants because the coarse-textured sand particles do not bind water molecules as tightly as the fine-textured clay-loam soils that support shortgrass prairie. The most conspicuous grasses and forbs of this season have root systems that are largely confined and most densely developed within the first two to three feet of the soil, an architecture that enables them to intercept moisture before it percolates down through the sandy soil.

As a scientist, Kelly Kindscher used sophisticated statistical analyses to validate his concept of tallgrass prairie plant guilds. But the numbers only affirmed what he already knew by immersion. As a boy, Kelly spent many summer days in the prairie pastures of his family's farm in the Republican River valley near Guide Rock, Nebraska, homesteaded by his great-grandparents in 1871. And in the spring of 1983, he and a friend *walked* from Kaw Point to the Rockies, a trek of nearly seven hundred miles. He undertook the odyssey "to broaden my perspective and deepen my knowledge of the Prairie Bioregion." In the process, he gained a deep feel for what he called "the spirit of the prairie."

If the natural treasures of the country of the Kaw are to be conserved in all their richness, we need more people like Kelly—sharp-eyed scientists and open-hearted ramblers.

PART II

The Membership

> "What good is it?" I answer: "It is good in that it plays a
> part in the web of life."
> *David Costello,* The Prairie World

The novels and short stories of Wendell Berry are set in the fictional town of Port William, a rural community in the hills of eastern Kentucky. The cast of characters that emerge, some admirable, some not, many humorous, comprise what Berry calls "The Membership." Whether they like it or not, the lives of these folks are deeply entwined, enriched as well as constrained by the web of relationships in their community. The country of the Kaw is likewise home to a diverse membership of plants and animals that demonstrate intriguing and often wondrous adaptations to the environment and to each other.

SEVEN

The Days of Manure River

> But now the main body of the herd was coming, one dark cover over all the upper forkings of the Republican River.
> *Mari Sandoz,* The Buffalo Hunters

Julian R. Fitch may have had a bit of the booster in him. Second lieutenant with the US Army Signal Corps, Fitch accompanied the Butterfield Surveying Expedition of 1865 to assess the pros and cons of a stagecoach route from Fort Leavenworth, Kansas, to Denver via the Smoky Hill River valley. His report seems a bit heavy on the pros, especially given the harrowing experiences of travelers who followed the Smoky Hill Trail to the Colorado gold fields in 1859 and 1860. He even proclaimed the region "the garden-spot and hunting-ground of America." But, to give Fitch the benefit of the doubt, he did witness the following: "Five miles west of Fort Ellsworth we were fairly in the buffalo range; for miles in every direction as far as the eye could see, the hills were black with those shaggy monsters of the prairie, grazing quietly upon the finest pasture in the world."

John Charles Frémont encountered a similar spectacle in 1843 about two hundred miles to the northwest on the divide between the Republican River and its tributary Beaver Creek; his journal entry for June 25 recorded "buffalo in great numbers, absolutely covering the face of the country."

It is safe to say the country of the Kaw was once prime bison pasture. The region may have been the greatest bison country in all of bison country.

Environmental historian Dan Flores has written extensively about bison ecology and the bison economy of the Great Plains. He uses the more scientifically correct *bison* rather than *buffalo* because true buffalo are species of Africa and Asia while the American "wild ox" (*Bos bison* or *Bison bison*) represents a different branch of the bovine family tree. Combining historical accounts with fancy ecological carrying capacity calculations, Flores estimates about thirty million of these animals lived on the plains during what he calls the great bison "efflorescence," a period of favorable climate that stretched from the 1500s into the 1800s.

Vast herds of bison were not everywhere present during these "grass-happy" days; their numbers and distribution ebbed and flowed locally and regionally in response to long-term changes in climate and short-term periods of extreme weather, particularly drought. But certain parts of the overall range of this species supported especially colossal concentrations.

In his 1877 book *Hunting Grounds of the Great West*, soldier-author Col. Richard Irving Dodge declared the region between the South Platte and Arkansas Rivers, "generally known as the Republican country," to be "the most prized feeding ground" in the southern bison range, adding, "It was the chosen home of the buffalo." His geographical description was likely behind the expression "Republican herd" used by Mari Sandoz in *The Buffalo Hunters* for one of four major concentrations of bison in the Great Plains. The Republican herd, Sandoz proclaimed, was "the greatest of all buffalo herds."

The eyewitness accounts of awestruck Euro-American travelers assert staggering numbers of bison out on the Republican and Smoky Hill Rivers. But testimony is also provided by the histories of the indigenous peoples of the region. The lifeways of nearly a dozen plains tribes were deeply entwined with the "Republican country." Taken together, it is clear that the country of the Kaw was, at least for a time, the most glorious bison range in the Great Plains.

Historical accounts of bison between Kansas City and the Flint Hills are few, with the famous exception of the first sighting of a "Buffalow" by the Lewis and Clark Expedition in the vicinity of Kaw Point in 1804. This scarcity was probably due to the century-plus presence of villages of the Kansa tribe along the lower Kaw and the arrival in the 1830s of forcibly relocated eastern tribes like the Delaware, Shawnee, and Potawatomi, who were trying to make a go of farming in the region. Beginning in the 1840s, heavy traffic along the Oregon Trail chased bison out of Big Blue River country as well.

We do know that the Flint Hills was once a bison stronghold. In 1724, a party of French explorers under the leadership of Etienne de Bourgmont departed the main Kansa village, at the time in what is now Doniphan County in northeast Kansas, and headed out to the Smoky Hills for a diplomatic parlay with the Plains Apache. They crossed the Kaw near the junction of Mill Creek and traveled southwest across the Flint Hills for a couple of days. The expedition journal records bison on the tallgrass uplands above Mill Creek "so numerous it is impossible to count them."

In 1806, Zebulon Pike and his exploring party traversed the Flint Hills a bit south of the route of the Bourgmont expedition and were "continually passing through large herds of buffalo, elk, and cabrie [pronghorn]." Pike noted that his Osage guides wanted to "destroy all the game they possibly could" as they crossed the Flint Hills because the area was the hunting ground of their enemies the Kansa.

By the middle part of the 1800s, bison herds in the Kaw watershed were concentrated to the west of the Flint Hills, out on the "upper branches"—the Republican and Smoky Hill Rivers. The reports of Fitch and Frémont bookend the heart of this bison hotspot. Fitch's encounter happened on the eastern edge of the Smoky Hills, where the prairie transitions from tallgrass to mixed-grass. Frémont was traversing the High Plains, where shortgrass prairie dominates the uplands.

No one knows for sure how long the Pawnee or their antecedents hunted bison in the central Great Plains, but Spanish and French documents from the early 1700s place the Pawnee in the region and indicate they were already a force to be reckoned with. There were at least two Pawnee villages on the Republican River in the late 1700s, but their main "towns" were situated on the Platte and Loup Rivers in what is today east-central Nebraska, a region archaeologist Waldo Wedel called the "Pawnee Crescent"—the heartland of their civilization.

The lifeway of the Pawnee combined river valley horticulture with twice-a-year communal bison hunts in which the entire village—every able-bodied man, woman, and child—would abandon their gardens and earth lodge dwellings and set out for the bison range.

While the timing varied considerably from year to year, the summer hunt usually got under way in June or early July and lasted until August or September, when they returned to their villages to harvest their corn, beans, and squash. The winter hunt commenced in late October or November with a return in March or April to prepare for planting. The objective, particularly for the villages of the South Band of the Pawnee, was the middle reaches of the Republican and Smoky Hill Rivers.

The Pawnee followed well-worn routes between the Platte and their hunting grounds to the south. The easternmost trace, the main "Pawnee Trail," left the Platte near today's Grand Island and headed south all the way to the "great bend" of the Arkansas River. The trail intersected the Republican River at a place the Pawnee called *Pa:hu:ru'*—"the rock that points the way"—commemorated today by the town of Guide Rock, Nebraska.

Depending on the location of the bison herds, the hunt might take place in the environs of the Republican or the village would keep moving south and then west into the watersheds of the Solomon, Saline, or Smoky Hill Rivers. The other major trail left the Platte valley about eighty miles to the west of Grand Island and followed Plum Creek and then Turkey Creek south, reaching the Republican River near today's Oxford, Nebraska.

At times these trails were a mass of humanity streaming to or from the hunting grounds. Missionary John Dunbar traveled with the South Band of the Pawnee on their winter hunt of 1834–1835. He estimated the procession of some two thousand Indians was about four miles long and observed, "This was much the largest company of horses, mules, asses, men, women, children and dogs, I had ever seen."

The Kansa also undertook communal bison hunts in the Kaw River watershed, as witnessed in 1724 by Frenchmen of the Bourgmont expedition. Bourgmont had arranged to tag along on a Kansa summer hunt out to the Smoky Hills region. A member of his party tallied the cavalcade as it passed and counted three hundred warriors, a handful of chiefs, three hundred women, and about five hundred children, plus at least three hundred dogs dragging the village's baggage by travois.

Anthropologists are tantalized by this documentation of the use of dogs as pack animals by a village tribe like the Kansa, which harkens back to the bison-shadowing "dog nomads" encountered by Coronado in 1541 on the southern Great Plains. That the communal bison hunt was practiced even before plains tribes acquired the horse hints at its antiquity.

The sight and sound of a train of three hundred pack dogs crossing the prairie is thrilling to imagine, but the thought of wrangling *five hundred children* on a multi-month, several-hundred-mile-long road trip is exhausting. Such communal bison hunts were massive, complicated undertakings that required extensive planning and ritualized preparations. And they were fraught with danger.

Evidence of Pawnee ascendancy in the region is found in the Lakota name for the Republican River, *Padani wakpah*, which translates to "Pawnee River. But another measure is the long list of their enemies. The seasonal bison hunts of the Pawnee put them at risk of collision with roaming parties of Kansa, Osage, Delaware, Cheyenne, Arapaho, Lakota, and other bitter rivals for this prime pasture.

Like a hawk taking wing from a cottonwood tree and getting mobbed by kingbirds, once the lumbering procession left the shelter of the Platte valley it might be set upon by any number of adversaries. That the Pawnee and Kansa would undertake such epic and danger-filled treks, two times a year, children in tow, is testimony to the tremendous abundance of bison in the country of the Kaw.

While the Pawnee had to travel at least a hundred miles to reach the region's best hunting grounds, the Cheyenne were living in the midst of it. In the early to mid-1800s, the Cheyenne held sway over a swath of the High Plains stretching from the North Platte River south to the Arkansas. True horse nomads, the Cheyenne pursued bison across this vast territory from spring through fall, but during the most brutal months of winter they hunkered down in scattered villages along streams that provided moving water, shelter from storms, and stands of cottonwood trees for heat and horse fodder.

The Cheyenne tribe at this time was organized into ten main bands, each with its customary wintering quarters. Three of these, the Ridge People, southern Suhtai, and Dog Soldiers, favored the country where the Republican, Solomon, Smoky Hill, and their tributaries ladder the landscape. The Cheyenne sometimes shared their winter camps with allies like the Arapaho and the Brulé and Oglala bands of the Lakota. The Dog Soldiers, feared warrior society of the Cheyenne, most often wintered in the Big Timbers area of the Smoky Hill.

In late spring or early summer, the dispersed Cheyenne bands would leave their winter camps and gather for a time in a single communal village. The upper reaches of Beaver Creek, in today's Rawlins County, Kansas, seems to have been a favored setting for these grand tribal encampments, presumably because it was the very best of their hunting grounds.

This is the same country Frémont was traversing in 1843 when he saw bison covering the face of the land. It must have been the sweet spot of sweet spots within the Cheyenne domain. Even today, minus the bison, this is magnificent country, with sweeping upland flats creased by bluff-bound streams lined with cottonwoods.

At these annual gatherings the Cheyenne would perform sacred tribal ceremonies and rituals, including, on occasion, the Sun Dance. Celebrated by many Great Plains tribes, the Sun Dance was intended to revive and renew the Earth and its inhabitants and to restore the harmony and vitality of life. As practiced by the Cheyenne, the Sun Dance included the erection of a medicine lodge structure in which secret rites were performed.

Once the ceremonies were concluded, the Cheyenne set out en masse on a communal bison hunt, the frame of the medicine lodge still standing in the now-deserted Sun Dance village, an offering to Maheo, the creator god. The Cheyenne exulted in their bison-rich empire and fought to the death to defend it, but they were humble enough to know they needed supernatural strength to hold on to it.

The most ironic evidence of the richness of the Kaw bison range is provided by the dimensions of lands allotted to the Kansa and Delaware tribes by the US government in the 1820s and 1830s. The Kansa reservation was a giant elongate rectangle thirty miles wide and two hundred miles long, stretching from near the junction of Soldier Creek with the Kaw out to the upper parts of the Saline and Solomon Rivers. A modern traveler driving Interstate 70 from Topeka to Hays would be covering the same ground. The Delaware lands were situated to the east of the Kansas but included the "Delaware Outlet," a strip of land ten miles wide and 150 miles long that ran along the north boundary of the Kansa reservation.

These tidy boxes, the contrivances of bureaucrats sitting in offices, were designed to provide the Kansa and Delaware with corridors to the High Plains bison range. But the outlets intersected the Pawnee Trail and emptied out into the heart of Dog Soldier country. Terrible bloodshed ensued.

As much as the country of the Kaw was the fiercely loved Middle-earth of these peoples, so too it was the land of their sunset. "The last buffalo hunt" is a sorrowful chapter in the history of many plains tribes.

The last tribal hunt of the Omaha people commenced in December 1876 when they departed their reservation in northeastern Nebraska for the bison range. The hunting party traveled for thirty days and covered four hundred miles before they finally encountered bison in early 1877 in the headwaters region of the Smoky Hill River in today's Wallace County, Kansas.

The last buffalo hunt of the Potawatomi tribe took place in 1877 when they traveled from their reservation in northeastern Kansas to the vicinity of today's WaKeeney, Kansas, again in the watershed of the Smoky Hill. En route, they passed by the recently founded colony of Nicodemus, a settlement of African American Exodusters from the South who had trekked to western Kansas for the opportunity to acquire public land. Oral traditions of both Nicodemus residents and the Potawatomi people tell of the Indians heroically sharing meat from their hunt with the struggling and starving settlers.

On August 5, 1873, a party of 350 Pawnee was wrapping up a communal bison hunt on the Republican River near today's Trenton, Nebraska, when they were ambushed by over fifteen hundred Lakota warriors of the Brulé and Oglala bands. The Lakota surprised the Pawnee and killed sixty-nine men, women, and children in a running battle through the loess canyons of the region. One of the Lakota leaders was Chief Spotted Tail of the Brulé, who less than a year earlier had entertained Grand Duke Alexis of Russia and an American entourage that included Buffalo Bill Cody and George Custer in a "royal buffalo hunt" on nearby Red Willow Creek.

The site of the attack on the Pawnee, considered the last battle between Indian tribes in America, is remembered as Massacre Canyon. The tragedy was also the last buffalo hunt for the Pawnee, who left Nebraska the next year for life on a government reservation in Oklahoma.

Clearly, there was something about the country of the Kaw that made it bison heaven.

The historical moment certainly played a role. Historian Elliott West called the headwaters region of the Republican and Smoky Hill Rivers the "middle country" of the nineteenth-century Great Plains. Situated east of the Rocky Mountains and wedged between the heavily trafficked valleys of the Platte and Arkansas Rivers, the region was essentially unseen and untouched by Euro-Americans prior to the Colorado Gold Rush of 1858–1861, when one of the routes to the gold fields ran along the Smoky Hill River.

Whites stayed away from the middle country for good reason. Veteran Indian trader Rufus Sage observed, "The region lying upon the head branches of the Kansas River is considered very dangerous—it being the war-ground of the Pawnees, Caws [Kansa], Cheyennes, Sioux, and Arapahos." Dan Flores speculates that fighting for control of the bison economy left a buffer zone at the boundaries where the warring tribes met, "occupied by neither side and only lightly hunted." This left bison and other game within the buffer zone relatively undisturbed, resulting in a buildup of herds.

Climate likely had a hand in it, as well. Paleoclimatologists have found evidence of an intense and persistent drought in the west-central Great Plains from 1845 to 1856 that matched or exceeded the severity of the historically documented disastrous droughts of the 1930s and 1950s. Centered along the western edge of the High Plains in Colorado, the drought would have dried up stream flow and withered plant growth, causing bison in the region to head east, into the headwaters and middle reaches of the Republican and Smoky Hill Rivers. So, at a time when warring Indian tribes and Euro-American incursions made the western reaches of the Kaw watershed a bison refuge, drought conditions to the west may have sent even more woolies into the scrum.

These factors no doubt played a role in concentrating bison in the country of the Kaw, but I believe it was the unique character of the land itself that made it an Eden for these creatures.

Renowned grassland ecologist Fredrick Albertson taught at Fort Hays State College (later Fort Hays State University) from 1918 until his death in 1961. His research efforts were focused on the mixed-grass prairie around the town of Hays and further afield in western Kansas. In 1937 he published a landmark paper, "Ecology of Mixed Prairie in West Central Kansas," which documented the surprising complexity and richness of the vegetation of the region.

The terrain around Hays is rocky and rolling, with broad, nearly level tablelands incised by numerous draws into hills and ravines with gently to steeply sloping sides. Such topography is typical of this part of the country of the Kaw, where the Republican, Solomon, Saline, Smoky Hill, and their many tributaries have scored and dissected the eastern edge of the High Plains surface.

This varied landscape results in a complex diversity of habitats, from dry windswept uplands with thin rocky soils to more sheltered vales with deeper soils. While the regional vegetation is broadly classified as mixed-grass prairie, Albertson's studies revealed a gradient of community types based on topographic position, with grasses and forbs typical of shortgrass prairie on the uplands, mixed-grass prairie on the hillsides and upper slopes of ravines, and tallgrass prairie on the valley floors and lower slopes of ravines.

To a hungry bison, the effect would be like a smorgasbord of greenery, stacked vertically. Compared to the Osage Cuestas with only tallgrass prairie, and the western High Plains with just shortgrass, the eastern edge of the High Plains offered a richer bill of prairie fare—mixed-grass *plus* tallgrass and shortgrass. The grasses and forbs that comprise each of these grassland types are most palatable and at their highest nutritional value at different times during the growing season, so their proximity would lengthen the grazing season locally, which would encourage the grazers to linger longer.

But a bison runs on water as well as grass, and a lot of bison would need a lot of water. Although saddled with a semiarid climate, a closer look at the landscape reveals unexpectedly rich resources.

Archaeologist Waldo Wedel knew his native Kansas and the central Great Plains like the back of his hand; he felt compelled to correct "the tendency among some observers to underestimate the High Plains region and its capacity to support native human populations." In a 1963 essay in the journal *American Antiquity*, the Smithsonian Institution scientist used historical accounts and early geological studies to demonstrate the nature and distribution of water resources during the heyday of the horse nomads. Wedel zeroed in on two unique attributes of the

High Plains landscape that sustained both bison and bison hunter—the Ogallala Aquifer and playa lakes.

The incised and dissected topography that enriches the plant palette in the western reaches of the Kaw also exposed water-bearing Ogallala Formation sediments in gullies, draws, and canyons, resulting in a surprising number of springs and seeps throughout the region prior to Euro-American settlement. One of the historical sources cited by Wedel was our Lieutenant Fitch, who noted numerous "fine springs" along the Smoky Hill River in 1865. Geological surveys as late as the 1890s confirmed strong and reliable springs on the upper tributaries of the Kaw in eastern Colorado, especially along the Arikaree and South Fork of the Republican.

Breaching of the Ogallala Aquifer in these valleys also allowed fingers of eastern deciduous forest and tallgrass prairie to extend into an environment that otherwise only supported shortgrass prairie. Although consisting of far fewer tree species than forests along the lower reaches of the Kaw, sometimes only cottonwood and willow, these narrow strips of timber offered bison shelter from the worst of High Plains weather plus grassy meadows for summer grazing and cottonwood twigs for winter browse.

These slim woodlands concentrated all manner of living things. Ethnographer George Bird Grinnell was told the story of some young Cheyenne men taking a swim in the Republican while agitated bison bulls were thrashing about in the timbered bottoms, shaking their heads in the thick underbrush. One big bull emerged and, feeling his oats, charged after the youths, his head wrapped in grapevines that trailed behind him like long ropes.

The intricately cut terrain of the upper reaches of the Kaw, particularly in the Republican River watershed, sets it apart from other streams that drain the central Great Plains. In 1849, Capt. L. C. Easton of the Quartermaster Department of the US Army was tasked with exploring the "Republican Fork" as a potential military supply route between Fort Leavenworth on the Missouri River and Fort Laramie on the North Platte in what is today Wyoming. The goal was to see whether the Republican provided advantages over the Platte River, which at that time was the primary route for travel to Oregon and California. Easton ended up endorsing the Platte as the best wagon road, but was still impressed with the Republican: "The Platte has but a few Creeks, flowing into it, while the Republican has an almost innumerable number."

Waterways are fewer in the far western reaches of the Kaw and are separated by extensive upland flats that appear all but featureless. But this part of the High Plains is dotted with playas—round, shallow basins that become evident only after they are filled by prolonged rains or a cloudburst or the melt of a spring snowstorm. While ephemeral and generally only a few acres in size, numerous historical accounts report dense concentrations of wildlife on these temporary ponds, especially bison. Wedel's essay includes a map that shows "many small ponds" about the High Plains headwaters of the Republican and Smoky Hill, and recent surveys confirm there are thousands of playas in the region.

I offer one final, earthy bit of evidence of the enormous presence of bison in our region. The old Pawnee name for the Republican River is *Kira ru tah* which, according to George Bird Grinnell, was derived from the Pawnee word for dung and translates to "Manure River." He was told the river was so named "because of the great numbers of buffalo which resorted to it, polluting the waters."

Concatenation refers to a chain of circumstances that coalesce to produce a unique, often unexpected, outcome. It seems that in the country of the Kaw an exceptional and perhaps unparalleled concatenation of climate, physiography, vegetation, water resources, and clashing human cultures intersected the biology of a big animal capable of becoming, in the words of Flores, "a weed species" and resulted in one of the greatest wildlife spectacles in recorded history.

EIGHT

O Elkader!

> It is said that the post office in Elkader was abandoned for this reason.
> *Theodore Scheffer,* The Prairie-Dog Situation in Kansas

Historical accounts from the country of the Kaw and elsewhere in the Great Plains reveal that wherever bison went, they trashed the place. Streams and playas were drained of water and left full of excrement. Swaths of grassland were grazed to the ground. The prairie sod itself was shredded by millions upon millions of sharp hooves, opening up the soil to erosion by wind and rain. Even more bare ground was exposed by the blissful wallowing of legions of itchy, thousand-pound beasts.

Bison are what ecologists call disturbance agents. Like wind and wildfire, drought and flood, the bison herds of yesteryear were a force that disrupted the environment and brought chaos into the local ecology. Bison are also considered ecosystem engineers, since their activities radically modify habitat, which in turn shapes community structure, which in turn influences the abundance and diversity of plant and animals in that community. Whatever label you use, the lives of just about every creature in the grassland world was under their sway.

We've already seen that there was a deep ecological intimacy between bison and shortgrass prairie, with the defining grasses blue grama and buffalograss actually prospering under bison foraging. But bison were gardeners as well as grazers.

Because these two grasses tend to grow in a dense turf, shortgrass prairie is not particularly rich in forbs. But flower patches often irrupt where the sod has been disturbed and bare ground exposed. Such places, which would have abounded during the zenith of the bison herds, have a higher diversity of forbs, especially plants with weedy tendencies like sunflowers and milkweeds. Increased diversity of plants translates into increased diversity of animals. The surge in biological diversity and food web complexity associated with bison disturbance is called an "ecological cascade."

The grandest designation applied to bison is that of "keystone species." Like the wedge-shaped block at the apex of a stone arch, a keystone species is the centerpiece of an ecological system, bearing the weight of the entire span. And, to keep the metaphor rolling, if the keystone drops out the whole arch and the structure it supports will come tumbling down.

It is no secret that earthshaking herds of bison are a thing of the past. But since bison were the grand mediators between grass and almost every other living thing in the Great Plains bioregion, it is possible we have lost or are losing plant and animal species whose life histories were intertwined with bison biology.

In a later chapter I discuss my work with the dwarf milkweed, a wildflower of the shortgrass prairie that appears to have been much more common in the 1800s. A few years after I did an extensive and mostly unsuccessful search for this rare plant, some relatively large populations were found on the Piñon Canyon Maneuver Site in southeastern Colorado, a training facility for the US Army. There, the milkweed often grows in areas where heavy traffic from tanks and other military vehicles has scuffed up the shortgrass turf and the prairie is on the mend.

It makes me wonder whether the ecology of this pretty little plant was once tied to the later stages of recovery from bison disturbance. The historical range of the plant certainly mirrors that of bison on the plains. Perhaps ecological specialization bound the dwarf milkweed to bulldozing bison, a once-dependable relationship that later cost the plant dearly.

Despite my suspicions about the dwarf milkweed, it doesn't appear that many plants or animals have life histories that were inextricably bound to bison. Yes, we've lost the gray wolves and ravens that shadowed the herds, but, as will be seen in another chapter, their demise was real-time collateral damage from market hunting operations. And not every creature that prospered from bison disturbance went down with the ship. Some, like buffalograss, dung beetles, and cowbirds, were able to transfer their ecological loyalties to Black Angus and Herefords.

There is another animal in the country of the Kaw with the ecological gravitas to be called a keystone species. Like bison in the past, prairie dogs played an enormous role in enhancing the biological diversity of the Great Plains. Their numbers are greatly reduced today, but there are still enough of them on the land for us to witness the amazing influence of a keystone species on the organization, stability, and functioning of the ecological system that orbits around them.

The prairie dog is the epitome of a disturbance agent. These colonial, burrowing rodents, technically ground squirrels, form extensive "towns" that provide tracts of highly altered terrain within the broader grassland matrix, terrain that attracts and supports a unique and diverse community of creatures. Zoologists recognize five different species of prairie dogs; ours is the black-tailed.

Viewed aboveground, a prairie dog town is dotted with a multitude of small hillocks and cones of packed soil, each marking the entrance of a tunnel that leads down into an elaborate subterranean system of chambers and secondary exits. The elevated mounds are built from soil brought up from the burrow or scraped up from the ground around the burrow entrance.

Some associates of prairie dog towns are there to forage on the sheer abundance of protein offered by such a concentrated multitude of rodents. Coyotes, swift fox, and badgers take prairie dogs by running them down or digging them up, while golden eagles, ferruginous hawks, and prairie falcons hunt them from the sky.

Of these predators, it is the swift fox and ferruginous hawks that are most closely bound to prairie dog towns and that have been the most severely impacted by declines in prairie dog abundance. The most specialized prairie dog predator is the black-footed ferret. This secretive, weasel-like creature, which seems to never have been very common, is now North America's most endangered mammal.

Other associates are there to take advantage of the subterranean architecture, including a number of reptiles and amphibians as well as small mammals. They can be found ducking into burrows to escape extreme temperatures or even using them for hibernating dens. But the most consistent denizens of their hidey-holes are prairie rattlers and western burrowing owls.

We'll see in a bit that prairie dogs have a serious public relations problem as relates to their town building activities. If the furry citizens were to suddenly become mindful of their image, they would no doubt downplay the rattlesnake info in their branding strategy. But, providing a stage for the antics of burrowing owls? Marketing genius!

There are nineteen different species of owls in North America, almost all of which are arboreal—*of the trees*. Only one is fossorial—*of the earth*. The burrowing owl nests belowground, placing its clutch of eggs in a burrow dug by a mammalian excavator. Given the abundance of accommodations available in a prairie dog town, a scan of a healthy colony will often reveal a scattering of the eight-to-ten-inch-tall, gnome-like birds.

The burrowing owl is one of the most winsome creatures of the western hemisphere, with an adorability factor exceeded only by that of otters. Humans have always found owls captivating, in part because they lead secretive lives under the cloak of darkness. Burrowing owls are unique in that they are active during daylight hours and can be seen going about their lives on the roofless pitch of the prairie dog town.

The owl's habit of staring you down from atop a prairie dog mound, head slightly tilted to the side and, if agitated, bobbing up and down, gives it a bit of a comical persona, what Nebraska ornithologist Paul Johnsgard characterized as "a spindly legged, feather-clad leprechaun still trying to recover from last night's hangover." Sure, many people find prairie dogs to be cute, but my guess is the burrowing owl has done more to earn tolerance for prairie dog towns than the rodents ever could have done on their own.

Burrowing owls depend on their hosts for more than just belowground accommodations. The approach of humans or other threatening creatures to the neighborhood triggers alarm calls from the prairie dog sentinels; the yips, chirps, and barks eventually spread throughout the colony, starting slowly then building to a chaotic crescendo as the dogs dash for their burrows. Burrowing owls, it turns out, "eavesdrop" on all of this chatter.

In a remarkable study set in prairie dog colonies in northeastern Colorado, researchers compared the response of nesting burrowing owls to broadcasted recordings of prairie dog alarm calls with their response to recordings of cattle mooing and an airplane engine. The owls significantly increased the intensity of their alert behaviors—head turning, stretching up tall, squatting low, vocalizing, bobbing—in response to the prairie dog alarm calls.

Following the Canadian River across what is today central Oklahoma, Edwin James of the Stephen H. Long Expedition of 1820 had a particularly sublime encounter with a prairie dog town. He paints a vivid picture in his account of the expedition:

> Here we passed a large and uncommonly beautiful village of the prairie marmots, covering an area of about a mile square.... The grass on this plain was fine, thick and close fed. As we approached, it happened to be covered with a herd of some thousands of bisons; on the left were a number of wild horses, and immediately before us twenty or thirty antelopes and as many deer. As it was near sunset the light fell obliquely upon the grass, giving an additional brilliancy to its dark verdure.

A good number of animals are drawn to prairie dog towns because of the unique vegetation associated with these colonies. Prairie dogs reduce the stature of grasses and forbs within the town by their own intensive grazing and clipping, and their burrowing activity and sheer amount of foot traffic creates large swaths of bare ground. The disturbed soil and open spaces are invaded by native species of weedy annual and perennial forbs that otherwise can't gain a foothold in the surrounding grassland matrix. These plants attract insects that gather pollen and nectar from flowers, herbivores that feed on foliage and flowers, and granivores that feed on seed, all of which serve as prey for a host of creatures farther up the food chain.

The presence of prairie dog colonies contributes to more diverse bird communities in the Great Plains. Certain grassland birds prefer the sparsely vegetated terrain of prairie dog towns to the surrounding shortgrass prairie matrix. The most notable of these is the mountain plover, an endangered bird that historically depended on prairie dog towns for breeding habitat. Much more common are horned larks, which hang out in prairie dog towns year-round but gather in especially big numbers during the winter, when their concentrated flocks attract the attention of prairie falcons. Meadowlarks, killdeer, mourning doves, and lark buntings also frequent prairie dog towns but are less tightly bound to them.

Another eddy in this ecological cascade involves harvester ants. Ecosystem engineers in their own right, harvester ants construct large, cone-shaped mounds over underground chambers in which they store seed collected from the environs of the colony. The cone, six to twelve inches tall and about two feet in diameter, is typically situated at the center of a circle of bare ground called a disk; the disk can be up to six feet in diameter and is kept free of vegetation by workers of the colony. Harvester ant mounds often occur in prairie dog towns, adding another scale of disturbance and patchiness to the already-jumbled tableau.

A study of the effects of prairie dogs on harvester ant colonies in southwestern Kansas found that, while the number and density of harvester ant mounds were about the same on prairie dog towns as in the surrounding shortgrass prairie, the species composition differed. The most frequent species on the prairie dog towns was the rough harvester ant, which happens to be the preferred fare of Texas horned lizards, commonly known as "horny toads."

So, the floristic bounty of the dog town sponsors seed-harvesting ants that sponsor ant-harvesting lizards. And, to close the loop, the seed-collecting forays of harvester ants help disperse seed, and their nest-building activities alter soil structure and create islands of increased nutrient density, which sponsors a greater diversity of grasses and forbs in prairie dog towns.

The forage quality of vegetation within a prairie dog town is often superior to that outside the colony, leading to preferential grazing by bison in the past and domestic cattle today. The churning and mixing of soil within a prairie dog colony improves soil moisture availability, making for more succulent plant tissues later in the growing season. Also, grazing of grasses by prairie dogs, particularly their preferred forage blue grama, causes the clipped plant to produce new growth, which often has a higher nutritional value (protein and nitrogen) than grasses growing outside the town.

The ecological cascade initiated by prairie dog activities provides refreshment for a wealth of other wildlife, including the most gorgeous mammal on the North American continent—the pronghorn.

What Edwin James and every other early Euro-American observer called antelope, modern-day zoologists refer to as pronghorn. Certainly, these graceful, speedy animals bring to mind the antelope and gazelles of Africa, but they are a distinctly North American species in a distinctly North American mammal family. I'm all for taxonomic accuracy, but I can't help but call them antelope, especially if there's not a wildlife biologist within earshot.

After bison, the pronghorn was the most visible big animal on the historic-era Great Plains. Environmental historian Dan Flores speculates they were at least as numerous as bison and, while the region was once the Great Bison Belt, "in truth it was just as much the Great Pronghorn Savanna."

But market hunting and conversion of rangeland to cropland precipitated a steep decline in their numbers, and by 1915 pronghorn had become extremely rare in the country of the Kaw, with only a few small bands persisting in sparsely populated counties in western Kansas and eastern Colorado. They have since made a bit of a comeback in our area, thanks in part to laws that initially prohibited and, later, limited hunting, plus the infusion of transplants from the northern and southern Great Plains in the 1960s and 1970s.

For this we can be thankful, because these animals are a thrill to behold. To witness a band of pronghorn skimming across the shortgrass prairie at high speed and then suddenly wheeling around and stopping to take a look at you is a breathtaking experience. Nineteenth-century Great Plains traveler W. W. H. Davis had it right: "The antelope is a most beautiful and graceful little animal, and, when running across the plains, has almost the appearance of a thing of air."

Pronghorn are renowned for their extraordinary speed (close to sixty miles per hour) and remarkable eyesight (they can detect movement up to four miles away). But they also have a famous weakness—their curiosity. Hunters past and present have taken advantage of the pronghorn's need to check out unrecognized objects by waving a cloth or some other article to draw the animal within shooting range.

Pronghorn also have a dietary weakness—scarlet globe-mallow. This showy wildflower produces an abundance of bright orangish-to-copper-red flowers, earning it the colorful name "cowboy's delight." Studies throughout the Great Plains have shown the late-spring-to-early-fall diet of pronghorn to consist largely of forbs (versus grasses and shrubs), with scarlet globe-mallow preferred above all others. Since scarlet globe-mallow prospers in the sort of disturbed habitat that abounds in a prairie dog town, pronghorn are drawn to these colonies like kids to an ice cream truck.

The "beautiful and interesting" scene captured for us by Edwin James represents a grazing mutualism hammered out over untold millennia. Prairie dogs set the table, bison grazed the grasses, and pronghorn browsed the forbs. It is noteworthy that the encounter happened on August 28, when the prairie is usually scorched by summer heat and drought, yet the vegetation in the prairie dog town was green and vigorous.

The extent of prairie dog colonies in the Great Plains in the early days of Euro-American exploration and settlement was staggering. James Mead, who hunted in the Smoky Hills in the mid-1800s, saw prairie dog towns that stretched for miles along the divide between the Saline and Solomon Rivers. In a paper given before the Kansas Academy of Science in 1888, A. B. Baker of WaKeeney described a dog town more than sixty miles in length in the valley of the Smoky Hill River. In a 1911 paper to the academy, Theodore Scheffer said that the largest prairie dog town in the state "extended almost continuously for about 125 miles along the Smoky Hill River and its tributaries."

An intriguing picture of the up-and-down fortunes of the prairie dog emerges from the country of the Kaw. While Mead encountered vast colonies in the Smoky Hills in the 1850s and 1860s, he later observed, "When the Buffalo left and ceased to tramp the ground, and the tall grass grew, the Prairie Dogs perished or disappeared." In his view, "The foot of the buffalo was necessary for their existence." For all their prowess as community builders, prairie dogs were essentially subcontractors to bison.

Elimination of bison herds was a severe blow to the prairie dog economy, but the little engineers staged a dramatic comeback when a new throng of even more effective disturbance agents arrived on the Great Plains—cattlemen and farmers.

Mead observed the resiliency of prairie dogs where the Chisholm Trail traversed now-bisonless central Kansas and the hooves of millions of Texas cattle wore ruts into the prairie as they were herded north to the railheads at Abilene and, later, Ellsworth. After operations ceased in 1885, the hundred-yard-wide trail "became a Dog town almost its entire length, the Dogs moving into it to secure the necessary conditions for their existence; to wit, hard tramped and almost bare ground."

But conversion of rangeland to cropland provided prairie dogs with an even greater recovery opportunity, particularly in the central Great Plains. When the settlement boom of 1885–1889 was ended by drought and a national financial crisis, farmers were forced off their lands, which left behind large swaths of abandoned farmland, conveniently broken in for the burrowing animals. These early settlers also unwittingly obliged prairie dogs by eliminating big numbers of their natural predators.

Scheffer, a zoologist at Kansas State Agricultural College, saw this play out dramatically in the valley of the Smoky Hill, where prairie dogs "infested" the region so thoroughly that settlers were actually forced to leave: "It is said that the post office in Elkader was abandoned for this reason—no patrons." Prairie dogs made a ghost town of Elkader.

The conflict between prairie dogs and humans has to do with the engineering and management of their colonies. The most obvious problem is the terrain created by prairie dog activity, a minefield of mounds and burrow entrances that invite crippling missteps by cattle and horses while also providing safe haven for rattlesnakes. Then there is the influence of prairie dogs on vegetation in and around the colony. Not only do the dogs graze and clip, they preferentially select blue grama, which puts them in competition with cattle. They also developed a taste for grain crops and garden vegetables.

As documented by David Wishart in *The Last Days of the Rainbelt*, favorable rains in 1891 and 1892 brought a second wave of settlement into eastern Colorado and adjacent southwest Nebraska and western Kansas, and these folks were not about to kowtow to the prairie dog. And so began the prairie dog wars. On the side of the settlers were state legislatures, township boards, college experiment stations, agents of the federal government, and other allies. On the side of the prairie dogs? No one.

In Kansas, legislation enacted in 1901 and bolstered in 1903 appropriated money for extermination of prairie dogs. Of various lethal control methods tested, poisoning proved to be the most expedient, either through toxin-laced grain baits that would be consumed or by fumigants that released poison gas when placed within the burrows. These same techniques are still in use today.

Of course, the impact of these "rodenticides" spirals well beyond the prairie dogs themselves and extends to any species that eats a poisoned dog or gets trapped in a fumigated burrow. And, on a much bigger scale, eliminating the keystone species was and is like taking a wrecking ball to the grand arch of ecosystem health.

In his 1911 paper, "The Prairie-Dog Situation in Kansas," Scheffer reported on the effects of this extermination campaign. After a survey of prairie dog territory in western Kansas, he concluded that "the prairie-dog is no longer a factor to be reckoned with by the farmer and the stockman," proclaiming, "the day of the dog is in its closing hours."

Not mentioned by Scheffer, maybe not even noticed by him, was the coincident disappearance of swift fox, black-footed ferrets, mountain plovers, burrowing owls, ferruginous hawks, and golden eagles from the prairie dog empire.

But the tenacious little animals were never completely whipped. Surveys in Kansas in 1991 and 1992 found sizable numbers of prairie dog towns in the western reaches of the Republican and Smoky Hill watersheds. At that time, the greatest estimated number were found in the Republican watershed, with sixty towns in Sherman County and fifty-two in Cheyenne County.

Prairie dog numbers were also high in the counties traversed by the Smoky Hill, particularly on flats adjacent to the river, where bluffs and badlands of Smoky Hill Chalk have weathered to yield the heavily textured soils favored by the burrowing animals. The largest prairie dog town in the state was found in this area, a sprawling, 583-acre colony in Scott County, the equivalent of nearly eight hundred football fields.

The persistence of prairie dog towns in the Smoky Hill River valley of western Kansas set the stage for a dramatic conservation saga, vividly portrayed by George Frazier in *The Last Wild Places of Kansas*. It began with the work of the Black-footed Ferret Recovery Implementation Team, assembled in 1996 by the US Fish and Wildlife Service and tasked with trying to reintroduce the near-extinct mammal to some of its former haunts. A wildlife-friendly Smoky Hill valley ranch couple got wind of the effort and wondered if the black-footed ferret might be brought back to Kansas. They reached out to like-minded neighbors on ranches in the area and, long story short, a complex of fifty-three thousand acres of privately owned rangeland was assembled, collectively hosting the largest prairie dog population left in the central Great Plains.

Despite tremendous opposition from local residents who viewed prairie dogs as vermin, reinforced by outside agitators with the same opinion of conservationists, twenty-four captive-raised black-footed ferrets were released onto the complex of ranches on December 18, 2007. Fifteen years later, ongoing tensions make it hard to get official information about subsequent releases or the current status of black-footed ferrets in the area—no one from the US Fish and Wildlife Service was willing to talk with me about it.

Ferrets aside, the combination of prairie dog abundance and remote and rugged country once made the Smoky Hill River valley of western Kansas a ferruginous hawk factory. A nine-year study (1979 through 1987) of the nesting ecology of ferruginous hawks in the area found 181 active nest sites, most placed on precipitous ledges weathered into outcrops and pinnacles of Smoky Hill Chalk. Sadly, a more recent survey (2019 and 2020) of these same productive nesting territories found a dramatic decrease in the number of active nests and the number of young raised.

The future of this magnificent and iconic Great Plains raptor—the largest hawk species in North America—depends on large areas of rangeland and proximity to its favorite prey, prairie dogs. The Smoky Valley Ranch is such a place. Located about ten miles from the site of old Elkader, this eighteen-thousand-acre preserve is owned and managed by The Nature Conservancy as both a working cattle ranch and safe haven for prairie dogs and their ecological entourage.

Following dissipation hastened by prairie dogs, the tiny trading center founded in 1887 as Elkader went through a couple of rounds of rebirth and renaming. In a stunning stroke of irony, one of these iterations was christened *Keystone*. If you don't believe me, check out *Kansas Place-Names* by John Rydjord or, better yet, stop in at the wonderful Keystone Gallery, located about halfway between Oakley and Scott City. Constructed of locally quarried Smoky Hill Chalk, the restored 1916 building, once home to a church that served the Keystone community, is now repurposed as a fossil museum and art gallery.

The notion that a single species could have a pivotal impact on the structure and health of an entire ecological system dawned on zoologist Robert Paine as he studied the influence of a species of starfish, the purple sea star, on the ecology of rocky intertidal pools on the northwest coast of North America. Paine coined the term "keystone species" in 1969, and the concept has since transformed the science of ecology and led to the recognition of keystones in an astounding variety of habitats around the world. Thankfully, one of Earth's most elegant expressions of keystone ecology can still be witnessed in the country of the Kaw.

NINE

Bird Sketches

> I am happiest in places, I should probably say habitats,
> where both Eastern and Western kingbirds are present.
> *Merrill Gilfillan,* Chokecherry Places:
> Essays from the High Plains

If, like poet Merrill Gilfillan, your well-being is bound up with kingbirds, then a summer sojourn in the country of the Kaw is just what the doctor ordered. Linking the eastern deciduous forest to the shortgrass prairie, our region hosts a living "double elephant folio" of the birds of America. There are eastern birds, western birds, grassland specialists, summer breeders, winter visitors, and a whole host of others just passing through. The total number is well over four hundred species, so we'll just pull out a few sketches.

"Dooryard" is an old-timey expression for that part of the farm not given over to crops and livestock. In *Birds of an Iowa Dooryard*, Althea Sherman distinguished the area around the farmhouse from the barnyard, stockyard, chickenyard, etc. From this cozy vantage point, Sherman carefully observed and wrote about the "home life" of birds that graced her family's farm in the early 1900s.

The subjects of her writings include chickadees, goldfinches, Baltimore orioles, grackles, blue jays, chipping sparrows, catbirds, house wrens, nuthatches, cardinals, robins, etc.—birds that "followed man to his home on the treeless prairies" and found that life with us isn't so bad. Today, we're inclined to speak of them as "backyard" birds, so familiar that they seem domesticated.

Dooryard or backyard, these birds are part of the fabric of human habitation throughout the country of the Kaw, and would be recognized by anyone living in the central and eastern United States. Most have woodland affinities and don't venture much out into the open. They are town birds, not country birds.

But out on the plains, one rowdy country bird has crashed the generally placid society of town birds. The western kingbird is an attractive species, with its pale gray head and back, bright yellow belly, and the classic regal bearing of the Tyrannidae family. But it is noisy to a fault and always seems worked up about something, scolding, bickering, and chattering away up in the treetops. Naturally birds of open country, western kingbirds have been making their way east for a number of years and have elbowed their way into the town bird guild of many communities.

Members of the Tyrannidae family, commonly called "tyrant flycatchers," are known for their pugnacious natures and willingness to take on larger creatures, verbally and physically. More than once I've seen a kingbird land on the back of a hawk and ride it for a while.

And they're not intimidated by our kind. The Blackfeet people of the northern plains gave a name to the kingbird that translates "stingy-with-his-berries," inspired by its habit of harassing women out gathering wild fruit. You can't help but admire the bird's moxie, even as you wish it would just shut up for a while. When we lived in Hesston, Kansas, there were days when kingbirds would make such a racket at dawn they would wake up our kids, way too early for a summer morning.

I know a Nebraska man, a retired physician, who is deeply involved in a number of conservation organizations. But you cannot schedule a meeting with him from the middle of April through the end of May, at least not in the morning. It is spring warbler migration season and, being in his late seventies, he is not sure how many he has left.

Warblers, technically wood-warblers, are small, active, brightly colored birds that feed primarily on insect larvae. Most North American warblers are Neotropical migrants; they winter in Mexico, Central America, and the West Indies and fly north to temperate areas to breed. Some travel all the way up into subarctic Canada, an astounding feat when you consider the diminutive size of these creatures.

Fifty-some warbler species migrate to or through the United States each year. Nearly forty of these could be encountered over a lifetime of birding in the country of the Kaw, although in a given year only about twenty species regularly make an appearance in our region. The majority of these would be found in the Kansas City area, at the fringes of the eastern deciduous forest and a bit farther west in wooded river valleys of the Kaw and its larger tributaries.

Each species of warbler has its own unique charm. Some have very distinctive coloration and patterning to their plumage and are easy to identify. Others are more subtle in coloration and require closer scrutiny, which can be challenging since warblers seldom stay still for very long. Many are exceedingly beautiful.

In addition to appearance, warblers exhibit notable differences in behavior, particularly in the way they forage for food. Black-and-white warblers work tree trunks like a nuthatch, probing the bark with their thin, sharp bills. Ovenbirds and orange-crowned warblers thrash about in the leaves on the forest floor. Blackpoll and chestnut-sided warblers forage at mid-levels in the canopy, while Blackburnian and Tennessee warblers focus on the treetops. Some, like the worm-loving warbler, prefer brushy margins of the woodland. The warbler way of divvying up the resources of a woodland is an elegant example of ecological partitioning.

A handful of warblers, like the American redstart, common yellowthroat, and yellow warbler, are summer residents and breeders in the country of the Kaw, but the majority are just passing through, some on very tight schedules. Warblers migrate primarily at night and feed during the day, particularly in the morning after a long flight in the dark. Some may hang around a patch of woods for only a day or two, lighting up local birding hotlines when word of their arrival gets out.

Once you fall for these birds, you must order your life accordingly. The migration patterns of warblers are complex, but after many decades of observations the general timing of the spring warbler season in a given area is fairly predictable, as is the expected arrival times of particular species. In our area, yellow-rumped warblers are usually back by mid-April,

Nashville warblers show up in late April, and chestnut-sided warblers in mid- to late May. Armed with this information, you can block out days on the calendar to deflect appointments and other obligations that might impinge on your warblering.

The quest for warblers requires heading into the woods in spring where, warblers or not, wondrous things always await. Like the mid-April day when I was hoping to catch the first of the year. I didn't see a single warbler, but the woods were jumping with ruby-crowned kinglets, brown creepers, and hermit thrushes. It was magical, especially since the kinglets, tiny things not a lot bigger than hummingbirds, were flitting about everywhere, even at my feet, unfazed by my presence. I went back the next day hoping for a repeat performance, but they had all moved on. A once-in-a-lifetime party with kinglets.

My mom, Mary Ann to her friends but "Mel" to her adoring family, had a wonderful sense of humor, and the smallest thing could set her to giggling. For some reason, she found the word *warbler* hilarious and, though she tried to restrain herself, the thought of me going out to *look* for warblers made her laugh until she would cry. Warblering has been likened to fly fishing, suggesting it is a more sophisticated and cerebral level of birding. My personal warbler count stands at twenty-two different species, but the memory of Mel's laughter reminds me that each new species sighted is a gift, not a cause for smugness.

People come to Nebraska from all over the world to witness the throngs of migrating sandhill cranes that gather on the Platte River in February and March. I try to make it out there each year to soak up the wild, noisy spectacle. But no less marvelous is the quiet passage of warblers up and down the continent.

Ponder the palm warbler! First encountered by a naturalist on a palm-graced Caribbean island, the splendid little birds refueling in our greening woodlands are making their way to spruce bogs up in the Canadian taiga. Captivating in beauty, astonishing in life history, warblers are a fleeting but sublime presence in the country of the Kaw.

In the generally drab world of sparrows, the Harris's sparrow has been called a "showstopper." Big—North America's largest sparrow—and boldly marked with black face, crown, and bib plus a large pink bill, they are striking enough that I did a double take when I first caught a glimpse of one in an Osage-orange hedgerow in Hesston.

Harris's sparrows have a certain cachet in the birding world. In addition to being "conspicuously handsome" they are a specialty of the central and southern plains, seldom seen to the east or the west. That makes them a prize of sorts, something that could compel a Massachusetts birder to jump in the car and drive to the Flint Hills. But they would need to schedule their trip between October and early May, because Harris's sparrows are only around here in the winter. They spend their summers and raise the next generation up north near Hudson's Bay.

There is something about these burly birds that stirs affection. An ornithologist writing in 1919 attributed their appeal to good looks and a "sprightly and vivacious manner." I developed a special fondness for the Harris's sparrow once I realized where it was discovered.

Thomas Nuttall was the first to publish a scientific description of this bird, which he encountered in 1834 while serving as a naturalist on an expedition led by Bostonian adventurer-entrepreneur Nathaniel Wyeth. The expedition departed Independence, Missouri, in April and followed what would later become the route of the Oregon Trail all the way to the Pacific Coast. "We first observed this species," Nuttall records, "a few miles to the west of Independence, in Missouri, towards the close of April."

The picture becomes clearer when you consult the journal of John Kirk Townsend, another naturalist on the Wyeth Expedition, who described departing Independence on April 28 and crossing "the *Big Blue* river at a shallow ford." We can't know exactly where the party flushed the sparrow that day, except that it was in the embrace of the Blue River of my boyhood, which warms my heart to no end.

Townsend said nothing about the sparrow in his journal but did note that as the party forded the Blue they encountered a number of yellow-headed blackbirds, stunning birds that, he added, "often alight upon the backs of our horses."

Lord have mercy!

Skylark is a luminous word that does duty as both noun and verb. It is the ancient name for *Aludia arvensis*, a bird of open country in Europe and parts of Asia, unassuming in plumage but spectacular in flight and song. Its use as a verb dates back to the days of sailing vessels when crewmen with time on their hands would *skylark*—scramble about up in the ship's rigging for fun. From these nautical origins, *skylarking* became shorthand for horseplay and roughhousing.

Such exuberant allusions are not surprising since, to the human eye, there is revelry in the skylarking of skylarks. So beloved are these birds that, like whippoorwills in the American South, skylarks are frequent figures in European song, folklore, and literature, including the works of English poet Gerard Manley Hopkins.

Singing on the wing is not unique to grassland birds, but skylarking is. The difference is the display component of the flight song. While goldfinches call to one another as they fly from sunflower to sunflower, the song of a true skylark is delivered as part of a uniquely structured aerial display. Skylarking seems to be a means of broadcasting song in open habitat where there are few raised perches.

Not every *lark* in name is a *skylark* indeed. Meadowlarks are beloved birds of our grasslands, and their song is as integral to prairie atmospherics as the wind. North America actually has two species of meadowlark, the eastern and the western, with the Flint Hills serving as something of a divide between the two.

Eastern and western meadowlarks are nearly identical in appearance but differ noticeably in their song, which is performed from atop a fence post or some other kind of elevated perch. Meadowlarks do deliver a quick flight call during the heat of breeding season, but it doesn't have the requisite aerial display to make it skylarking.

Six Great Plains songbirds are bona fide skylarks—chestnut-collared longspur, McCown's longspur, Sprague's pipit, lark bunting, Cassin's sparrow, and horned lark. You can witness performances of the last three in the country of the Kaw.

Lark buntings are large, stout-bodied sparrows with conspicuous patches of white on their wings. The streaky, dried-grass coloration of females and immature males is unremarkable, but males in breeding plumage are striking—jet black from head to tail except for the white of wing patches and tail corners.

These "eminently gregarious" birds winter in northern Mexico and the southwestern United States; they head north to the Great Plains to breed and arrive in the country of the Kaw by late April. Lark buntings have a preference for shortgrass prairie and can be locally abundant in good habitat. They are such a familiar presence on the plains of eastern Colorado that in 1931 the lark bunting was named the state bird, a controversial pick when there were plenty of compelling mountain birds to choose from.

Males begin skylarking soon after arrival on breeding grounds and keep it up for several weeks. Springing up from the ground or a low shrub, singing as they ascend twenty feet or more in the air, they slowly float back down, still singing away, on wings held in a steep V pattern. The floating part of this captivating performance has been called the "butterfly display."

Lark buntings are not strongly tied to particular breeding areas and may be locally common one year and absent the next. Fortunate is the person who stumbles upon a place where a flock of lark buntings has set up shop for the breeding season. But brace yourself. The relentless chase of males after mates and rivals, wing bars flashing, joined with fervent singing and eruptions of skylarking, is mesmerizing and can make the watcher wobbly.

Cassin's sparrows are birds of shrubby habitat, mostly in the southern Great Plains, and your best chance of seeing one in the country of the Kaw is in the sandsage prairies of the Wray Dune Field in northeastern Colorado and southwestern Nebraska. Their skylarking is similar to that of lark buntings. Males launch from a sagebrush shrub to a height of up to thirty feet and then glide flat-winged, head up and tail spread, to another shrub, singing all the way. In their delightfully personal book, *Oklahoma Bird Life*, husband and wife authors Fredrick and Marguerite Baumgartner describe the song as "a tinkling sound like the chimes of a tiny music box." I detect snatches of meadowlark.

Horned larks are the most common birds of the shortgrass prairie but are the hardest to get to know. With plumage the color of dust and a restive nature that keeps them always flitting away from you, they seem intent on dodging the spotlight. Yet, as is sometimes the case with introverts, horned larks are capable of short bursts of extroversion. Ornithologist Gayle Pickwell called the flight song of the horned lark "one of the most remarkable activities in the bird world."

The aerial display begins with the skylarking male working his way up to the desired altitude, up to five hundred feet above the ground, so high the bird all but disappears from sight. If the wind is calm, this is accomplished by ascending in a series of large circles, like the arcs of a helix. On blustery days, which is typically the case on the western plains, he faces directly into the wind and climbs in a series of steps interspersed with floating pauses, like scaling an invisible ladder.

Once the desired height is reached, the bird maintains altitude by spreading his wings and tail, fluttering as needed to hold the zenith, and starts to sing, pouring forth a rapid jumble of twitters and trills.

The male stays aloft for a minute or two, sometimes longer, and then comes the stunning climax. Closing his wings, he just drops like a rock, plummeting headfirst back to earth, and breaks the descent at the last moment by spreading his wings.

Pickwell, who was captivated by horned larks as a Nebraska farm boy and studied them in the 1920s for his master's thesis at Cornell, timed the drop and found it lasted from four to six seconds. He called this aspect of the horned lark flight song "the most thrilling dive ever indulged in by living creature." I have twice witnessed a skylarking horned lark, and the dive left me thunderstruck.

Willa Cather famously wrote, "Between that earth and that sky I felt erased, blotted out." A lark bunting would find such sentiment puzzling. We humans may be unnerved by the sweep of the plains, but this is the lark bunting's happy place, the corner of the cosmos where a skylark can skylark. Crying, in the words of poet Hopkins, *What I do is me: for that I came.*

The country of the Kaw provides a corridor of potential overlap between several pairings of eastern and western birds. The eastern member of these duos usually favors oak-hickory forest, while the westerner occurs in sheltered ravines and canyons or cottonwood-shaded streams. Along with kingbirds, these include phoebes (eastern and Say's), wood-pewees (eastern and western), orioles (Baltimore and Bullock's), buntings (indigo and Lazuli), flickers (yellow-shafted and red-shafted), and grosbeaks (rose-breasted and black-headed). A number of other birds of eastern or western affinity lack such counterparts but reach the limits of their continental distributions somewhere within the reaches of the Kaw.

Once when traveling across the divide between the Smoky Hill and Republican Rivers in eastern Colorado, I came across a jarring avian mashup. I had pulled off the state highway to check out a historical marker in a roadside rest area planted with Siberian elms and cedars. Western meadowlarks were singing away out on the prairie and western kingbirds were fussing up in the treetops. Adding to the occidental vibe was a pair of Bullock's orioles ferrying insects to the chirping young in their hanging-basket nest.

But the loudest sound in the otherwise peaceful scene was the voice of a mockingbird, doing pitch-perfect renditions of meadowlark song and kingbird chatter. I had always associated mockingbirds with the South, where they mimic cardinals and Carolina wrens from atop magnolias. Not only was this High Plains mocker ridiculously far afield, it was also picking on the locals like a local.

Most of the raptors hunting the country of the Kaw can be found on the bird lists of eastern states like Indiana or Ohio, even Georgia. But a handful—three hawks, two falcons, and an eagle—prefer to work the big lonely skies over the plains.

The Swainson's hawk, already mentioned, is the most frequently encountered summertime hawk on the western plains. Migrating each year between the grasslands of North and South America, these graceful birds, "one of the most striking and handsome buteos found anywhere in the world," start arriving in our area in early April and are generally gone by the middle of October.

Swainson's hawks hunt mice, ground squirrels, and other small vertebrate prey during the breeding season; when done feeding their young, they make a dietary switch to large insects, especially grasshoppers.

When enormous infestations of the High Plains grasshopper swept across the plains during the Dust Bowl days of the 1930s, Swainson's hawks came in by the thousands to gorge on the bonanza. Albert Munsell Brooking of Hastings, Nebraska, observed a similar spectacle during a grasshopper outbreak in 1896: "I shall always believe that the Swainson's Hawk saved the day, for right with the 'hoppers' came the hawks, literally thousands of them."

The Great Plains is to rough-legged hawks what the Gulf Coast is to midwesterners—a place to get away from winter. These Arctic raptors arrive in the country of the Kaw in late September and early October, about the time the Swainson's hawks are clearing out, and head back north in March and April.

The bane of lemmings during Arctic summer, wintering rough-legs prey on small grassland mammals like mice and voles. Rough-legs are particularly adept at working air currents, a skill that benefits them when hunting in open country with few perches. Their signature technique is *kiting*, where, with wings and tail at full spread, the bird hangs motionless in the wind while scanning the ground below.

The ferruginous hawk is the largest hawk in North America and, with a wingspan of about four-and-a-half feet, it can take larger prey like prairie dogs and pocket gophers. The common name refers to the rufous or rusty-brown markings, the scientific name, *Buteo regalis*, to its imposing size.

Unlike Swainson's and rough-legs, ferruginous hawks are present year-round on the plains. Yet they are seen less often than the other two, in part because of their preference for remote country during breeding season, but also because efforts to reduce or eliminate prairie dog towns has taken a toll on their numbers. Your best chance of seeing a ferruginous hawk in all of the central Great Plains is in and about the badlands and canyons of the Smoky Hill River in western Kansas.

Prairie falcons and merlins drift into the country of the Kaw in the fall, prairie falcons from the west, merlins from the north. Both of these skyrockets are drawn to the plains by the huge numbers of horned larks that congregate here in winter, tumbling across rangeland and wheat fields in massive flocks that often include Lapland longspurs.

Prairie falcons are the larger of the two, about the size of a peregrine; they nest in rugged terrain in the mountains and canyonlands of western North America. Merlins are smaller falcons, just a bit bigger than the familiar American kestrel; they occur throughout the northern hemisphere and nest primarily in coniferous forests. Ornithologists recognize nine subspecies across the global distribution of merlins, including one whose breeding territory is concentrated in the prairie provinces of Canada. The wintering range of this "prairie merlin" seems to be centered on the plains of eastern Colorado.

With a wingspan of nearly seven feet, the golden eagle deserves all the superlatives you can muster. Their great size makes them unmistakable, whether they are circling above or perched on some elevation, and when the sun hits the golden wash on their head and nape, it will make your knees buckle.

An encounter with one of these magnificent birds feels like an audience granted with royalty. I had the privilege of being buzzed by a golden eagle while I was botanizing along the South Fork of the Republican in eastern Colorado. The eagle was tearing along the edge of the river's bluffs, probably in pursuit of a jackrabbit, and I don't think it ever saw me on the slope below. What sticks with me now, some thirty years later, is the sound of the wind rushing over those wings.

It is pretty hard to top an experience with an eagle, but there was that incident in Ash Hollow. . . .

Ash Hollow is a patch of rough country in western Nebraska and a legendary landmark on the Oregon Trail. Most nineteenth-century travelers reaching this sheltered oasis of trees and springs had spent at least a few days in the flowery tallgrass prairies of the country of the Kaw as they trekked toward the Platte River. Such was the Pattison family of Sparta, Illinois, who were encamped on the Little Blue River on June 3, 1849, probably near today's Hanover, Kansas, and arrived at Ash Hollow on June 18.

Rachel Pattison died here the very next day, a victim of cholera. Only eighteen, she had wed Nathan Pattison just two months earlier. The entry in Nathan's spare diary simply records, "on the morning of the 19 Rachel was taken sick and dyed that night."

Nathan buried Rachel at the base of a hill that, on its other side, joins a long line of bluffs overlooking the valley of the North Platte River. I have climbed this rise, past Rachel's grave, a number of times. The hill crests in a handsome stand of prairie, thick with threadleaf sedge, and the view is expansive and powerful.

Once, on a blustery, hard-edged day in March, I reached the top to discover a raptor working the bluff line to my right. It was drifting toward me and before long I was able to make out its markings—heavily feathered legs, a patch of black at the wing wrist. It was a rough-legged hawk, nearing the end of its winter vacation in Nebraska.

Suddenly, down the bluffs to my left, I glimpsed another! I followed them both as best I could, each seemingly lost in its own private aerials

—wheeling, plunging, hanging in midair. Then, gradually, I began to perceive an aim. Slowly, exquisitely, they were making their way toward each other, working the roaring updrafts until they were right in front of me, floating, side by side, just off the edge of the bluffs.

Their union held just long enough for me to both grasp and doubt what I was seeing. Then a blast of wind swept them away, out of my sight. They were gone in an instant but I knew where they were headed—north, to the Arctic, to raise more rough-legs.

My heart broke, for Nathan Pattison.

In *Rivers & Birds*, Merrill Gilfillan writes, "All birds acquire rigorously subjective associations for people who notice them in any detail." He goes on to describe his deep connection with the yellow-breasted chat, a type of warbler, which grew out of seeking healing for a sick friend through a Lakota sweat lodge ceremony on the Pine Ridge Indian Reservation in South Dakota. The ritual lasted four days, and each night they emerged from the sweat lodge to hear chats singing away down along Wounded Knee Creek. "Accordingly," he writes, "they now carry heavy freight."

Birds are naturally emblems of joy (cardinal song on a snowy February morn!), but we also turn to them in our sorrows. Take for example the discovery of a "complete skeleton of an eagle" buried with a young child in a grave on a hill above an early nineteenth-century Pawnee village site on the Republican River. Describing the finding in a 1927 article, archaeologist Asa T. Hill wrote that the eagle "rested on the child's breast," adding, "showing how high was the esteem of this little one and how deep the love of those who laid the child away." It may not be culturally appropriate to speculate on the meaning ascribed to the eagle, but there can be no doubt that it had meaning, and that meaning was thunderous.

Or consider Menno Schmidt, late of Marion County, Kansas. I had only two conversations with the man, and both times I saw tears in his eyes. The first was when he told me the story of his daughter, a beautiful, world-ahead-of-her teenager when she suffered a traumatic brain injury in a car wreck. The second was when he spoke of cedar waxwings.

Like most Mennonite farmers, Menno's life was entwined with that of his crops. Bushels were the measure of the man, in all the ways a man might weigh his worth. *Nature* was what happened in the wheat field. But his beloved daughter's accident, and the unending heartbreak of her disability, led him to see birds. And in that seeing, there was some measure of healing.

Ever the husbandman, Menno began to cultivate birds: feeders were set out, and trees were planted; field corners were turned to wildlife cover; parcels of cropland were restored to wetland. His work was rewarded and bird books were needed. Among his new acquaintances was a sleek, crested, fawn-colored thing of piercing beauty.

Our waxwing discourse happened on a walk, as we swapped stories of bird sightings. We had finished with scissor-tailed flycatchers, summertime residents in the area, when cedar waxwings came up. There was a lag in the conversation and I caught the far-off look, and saw the moisture in his eyes. I changed the subject.

I'm glad birds were there for Menno when he needed them. They've been there for me.

TEN

Surprised by Shorbs

> The upland sandpipers reassure one another as they pass
> on quivering wings in the dark above my head, "the Flint
> Hills tonight, the pampas by the weekend."
> *John Zimmerman,* Cheyenne Bottoms: Wetland in Jeopardy

I have to confess to queasiness about wetlands—those places of sogginess, thick vegetation, swarms of bugs, and musky air. But, if I'm honest, there is something more to my aversion than these annoyances, something that sets off alarms deep in my limbic system.

Water moccasins.

Water moccasins are big venomous snakes that lurk with their bad tempers in swamps, ponds, sloughs, lagoons, and marshes (read *wetlands!*) in the southeastern United States.

The crazy thing is, except for two brief years in southern Illinois, I've never lived where water moccasins live. As a botanist living and working in the Great Plains, I've spent lots of time in rattlesnake country, and have had several close encounters; but rattlers don't mess with your mind. Unlike water moccasins, rattlesnakes don't drop from trees, open the door of your car, hide under the seat while you drive home, let themselves into your house, and wait for you to fall asleep. . . .

I was visiting a nature center in Mississippi a few years ago and, making conversation with my hosts, casually mentioned my irrational fear of water moccasins. The southerners found this quite amusing and redirected our tour to a pond known for its resident population. There they were, maybe a dozen big uglies lolling about on the bank and draped over tree roots, with doubtless hundreds more up in the trees. Even now, the thought of that pond makes my throat tighten.

So, it shouldn't be surprising that I was slow to warm up to the Rainwater Basin.

The Rainwater Basin is a flat to gently rolling plain of about six thousand square miles in south-central Nebraska. Lying just south of the Platte River, the region encompasses most of the headwaters of the Big Blue and Little Blue Rivers. Prior to Euro-American settlement, the Rainwater Basin was a broad expanse of tallgrass prairie grading into mixed-grass prairie on its western edge. Widely scattered throughout this region of deep loess soils are shallow wetlands that form where clay-bottomed depressions catch and hold precipitation. These wetlands, technically playas, are filled temporarily by rainwater and snowmelt and range in size from less than one acre to over a thousand acres.

I had been driving across the Rainwater Basin, north to south, east to west, for over three decades before I finally decided to take a closer look. It is an easy place to hurry across. Even a sympathetic account of the Rainwater Basin called it a region of "indistinct topography" and declared the wetlands, the key distinguishing feature of the landscape, to be of "unexceptional character."

As if this isn't bad enough, there is very little natural habitat still to be found within the nearly four million acres of the Rainwater Basin. It is estimated that at the time of settlement more than eleven thousand wetlands were spread across the region. Today, over 90 percent of the original wetlands have been destroyed by draining, and the landscape has been almost entirely converted to corn and soybeans. The wetlands that do remain, totaling about thirty-four thousand acres, have been degraded to varying degrees.

And yet, each spring, the shorebirds come pouring in.

The wetlands of the Rainwater Basin have long been known as a great place for waterfowl, with as many as fourteen million spring-migrating ducks and geese visiting the region annually. The Arctic-nesting snow goose can be spectacularly abundant in the Rainwater Basin from late February into March, something I witnessed in 2019 just a few days after the central Great Plains was clobbered by what meteorologists call a bomb cyclone, an explosive weather event that is essentially a winter hurricane. The storm, which hit Nebraska on March 14, caused catastrophic flooding of rivers in the central and eastern parts of the state and turned many Rainwater Basin wetlands into lakes.

I was driving on country roads south of Kearney a couple of days after the storm when I came across a thoroughly inundated wetland teeming with snow geese. I got out of my truck and was standing there, gawking at the sight, when I glimpsed what I thought was an approaching bank of storm clouds. But the rolling billows consisted of even more snow geese, hell-bent on joining the party on this three-day-old lake.

A group of birders from the Audubon Society's Rowe Sanctuary on the Platte River came upon the scene later in the day and estimated forty acres of snow geese consisting of 1.7 million birds! The sight was incredible but so was the roar of a million-plus honkers having a great day and telling each other about it. It sounded like Nebraska's Memorial Stadium on a Saturday with Oklahoma in the red zone. The bomb cyclone had spawned a snow goose apocalypse.

It wasn't until the 1990s that ornithologists and wildlife biologists began to fully realize the importance of the Rainwater Basin to migrating shorebirds. *Shorebird* is a catchall name for birds that typically associate with shorelines, especially coastlines. All shorebirds are migratory, and some undertake annual journeys of staggering magnitude between southern wintering areas and northern breeding grounds. A good number pass through the mid-continent during migration and often congregate at points along the way to rest and refuel, much to the delight of land-locked birders.

That lots of shorebirds migrate through the Great Plains in spring was evident from the well-known concentrations at Cheyenne Bottoms in central Kansas, the largest interior marsh in the United States, and the salt flats of Quivira National Wildlife Refuge just to the south. But the dispersed nature of wetlands in the Rainwater Basin masked its importance as a stopover area.

Thanks to the pioneering survey work of Nebraska ornithologist Joel Jorgensen, we now know that in most years two to three hundred thousand shorebirds representing nearly forty different species visit the Rainwater Basin in April and May on their way to breeding grounds in the

North, with up to six hundred thousand birds in years of favorable conditions. Accordingly, in 2009 the Rainwater Basin was designated a Landscape of Hemispheric Importance by the Western Hemisphere Shorebird Reserve Network.

I knew none of this when on an April day in 2018 I set out from my home in Lincoln in hope of catching up with a group from the Audubon Society of Omaha on a field trip to "the basins." I turned onto the gravel road leading to a well-known wildlife production area in Seward County and found them—a small caravan of Subarus, Priuses, and other standard birder vehicles stopped on the road, the occupants out scanning the wetlands with binoculars and spotting scopes.

Something in me snapped that day. With the help of these knowledgeable and gracious folks, I made the acquaintance of the dunlin, semipalmated plover, Wilson's snipe, Hudsonian godwit, marbled godwit, greater yellowlegs, long-billed dowitcher, and pectoral sandpiper. I went back out on my own several more times that spring and added a number of other shorebirds to my list. Then I started to read up on these birds and discovered that each has its own unique and often extraordinary migration story.

That all this biological theater takes place on such an unremarkable stage far removed from any ocean and less than an hour from my front door boggled my mind and stirred my soul. I fell in love with the Rainwater Basin.

Shorebird identification can be daunting, especially sorting out the diminutive, indistinctly marked kinds that skitter around in mixed flocks. Even seasoned birders sometimes give up and just write "peeps" in their field notes.

Thankfully, a number of "shorbs," as the aficionados call them, are distinctive in size, profile, and coloration and are not too hard to identify. First on this list are the large, lanky, and wonderfully beautiful American avocet and black-necked stilt, which wade their way through wetlands with a splendid grace. Then there are medium-sized, rather chunky, long-billed species like godwits, dowitchers, and Wilson's snipe. Finally, there are smaller but distinctively marked or proportioned species like

American golden plover, killdeer, dunlin, stilt sandpiper, and Wilson's phalarope.

As for the peeps, so named for their faint but persistent little callings, the mixed flocks of sparrow-sized shorebirds working Rainwater Basin mudflats in spring often include least, semipalmated, and white-rumped sandpipers. Western sandpipers are more common in late summer gatherings of south-bound peeps.

The feeding behavior of shorebirds can be a helpful aid to their identification, like the zany spinning swim of Wilson's phalaropes, which creates a small vortex of water that concentrates prey kicked up from the mud by its feet, or the rapid up-and-down "sewing machine" technique of dowitchers probing the mud with their bills. Then there are habitat associations—stilt sandpipers feed in shallow open water, semipalmated plovers prefer recently exposed mudflats, and pectoral sandpipers like the grassy edges of wetlands. There are also more subtle clues, like how lesser yellowlegs move methodically across submerged mudflats while greater yellowlegs feed by "dashing about."

These feeding patterns ultimately reflect the unique dietary preferences and foraging strategy of each species, a differentiation that helps partition wetland fare among potential competitors.

What's on the menu? The main entrée, the house specialty, is bloodworms, the larvae of midges, those small flies that swarm in annoying clouds around your head in summer. Kansas ornithologist John Zimmerman called bloodworms the "cornerstone" of the wetland ecosystem in the Cheyenne Bottoms and the "fuel" of intercontinental shorebird journeys. Bloodworms are thickest in the muck and sludge of submerged mudflats, where they glean nutrition from decaying organic matter. Shorebirds also comb the wetlands for aquatic insects, small crustaceans, mollusks, worms, and some vegetable matter.

It is heartening that these shorebirds find sustenance in the less-than-pristine wetlands of the Rainwater Basin. But a handful of their kin have made even greater accommodation to the heavy hand of humanity.

Unlike most shorebirds, American golden plovers, Baird's sandpipers, buff-breasted sandpipers, and upland sandpipers spend more time on upland sites than in wetland habitat. Prior to the near complete conversion of the Rainwater Basin to farm ground, this upland habitat was cloaked with prairie, and these birds mainly foraged in areas where recent burns or heavy grazing kept the vegetation relatively short. Now this terrain is almost entirely planted to corn or soybeans.

Still, these resilient birds continue to show up each spring and utilize the Rainwater Basin as stopover habitat on their northward migrations. While they do visit wetlands to bathe and drink, they are mostly found in dry, recently tilled agricultural fields. The ornithologists who study them call them "agripipers."

The buff-breasted sandpiper is the poster child for the agripipers. This dove-sized bird looks like a typical sandpiper but behaves very differently than most, not only in its preference for dry upland habitat but in aspects of its social behavior, which includes lekking, a mating system where males establish a site on the breeding grounds and put on elaborate courtship displays for the ladies.

As a species, buff-breasted sandpipers are considered highly imperiled and of significant conservation concern. Yet they are somewhat common in the eastern Rainwater Basin in late April to mid-May, as first reported in the 1970s by bird-loving York County farmer Lee Morris, who would watch them flocking to newly plowed ground from the seat of his tractor. Morris's observations led to systematic surveys of the Rainwater Basin by Joel Jorgensen, who confirmed that "exceptionally high densities" of buff-breasted sandpipers visit the region in spring.

It turns out that a large proportion of the world's population of buff-breasted sandpipers, maybe 90 percent, stop in the Rainwater Basin during spring migration to the Arctic; the region is critical to the species' survival. Ironically, it was the intimate connection between the Rainwater Basin and this agripiper that led to recognition of the region as a Landscape of Hemispheric Importance to shorebirds.

Drs. John McCarty and LaReesa Wolfenbarger, both professors of biology at the University of Nebraska-Omaha, have been studying Rainwater Basin agripipers for many years, with a special emphasis on buff-breasted sandpipers. They let me tag along one day during their annual spring migration survey. We visited thirty-five sites where "buffies" (that's what they call them) had been sighted in the past and found birds at three of the sites.

The best one had fifty-nine birds, most of them very active, chasing each other and exhibiting courtship displays, which involve flashing of the white underside of a single wing and a double-wing cupping embrace. Migratory birds generally refrain from this kind of energy-sapping behavior until they reach their breeding grounds. The exuberance of social interactions displayed by buff-breasted sandpipers in the Rainwater Basin raises the possibility that spring migratory stopover in the region may play an important role in group or pair bond formation, another facet of the deep relationship between this intriguing bird and this tattered landscape.

Almost all of the nearly forty species of shorebirds that visit the Rainwater Basin in spring are passing through and not looking to spend the breeding season with us. Some, like avocets, willets, and Wilson's phalaropes, may only continue as far north as the alkaline lakes of the Nebraska Sandhills, but most still have a long journey ahead of them.

Marbled and Hudsonian godwits are both regular springtime visitors to Rainwater Basin wetlands. They are easily distinguished from other shorebirds by their larger size and long, slightly upswept bills. Marbled godwits are a little bigger than Hudsonians and, when in flight, show a beautiful cinnamon wash on their underwings.

"Godwit" is one of the great names in the bird world—I find it hard to say without cracking a smile. Its origins are obscure but may have come from the Old English *god wiht* or "good thing," perhaps a reference to the tastiness of one of the two European species. This seems plausible since godwits are a bit huskier than most shorebirds.

Flights of both marbled and Hudsonian godwits begin arriving in the Rainwater Basin in April and it is not unusual to find the two species feeding together. But these overlapping flocks are on very different journeys. The marbled godwits wintered in coastal areas of the southern United States and Mexico and are headed to prairie wetlands in the Dakotas, Montana, and southern Canada. The Hudsonians started their trek at the southern end of South America and will continue on all the way to Arctic and sub-Arctic Alaska and Canada.

Hudsonian godwits are not the only shorebirds to visit the Rainwater Basin en route from the far end of South America to the Arctic hinterlands of North America. American golden plover, Baird's sandpiper, and buff-breasted sandpiper also migrate the entire length of the Western Hemisphere, two times a year. But the journey of the Hudsonian godwit is the most amazing of them all.

With the breeding season behind them, Hudsonian godwits from across Alaska and northern Canada head for James Bay, the southeastern thumb of their namesake Hudson Bay, where they gather by the thousands in late July and early August in preparation for return to their wintering range. They then launch out on a nonstop, three-thousand-mile flight to South America, most of it over the Atlantic Ocean.

While this is an astounding feat, the north-bound spring migration of Hudsonian godwits, even more epic, pulls the Rainwater Basin into the story. According to recent research by ornithologist Nathan Senner of the University of South Carolina, at least some of the Hudsonians arriving in the Rainwater Basin in spring just made a six-thousand-mile, *nonstop* flight from the coast of south-central Chile to get here. That is about seven days of flying, with no food, water, or rest, this time over the Pacific Ocean.

I appreciate Dr. Senner's research, but all of this gives me vertigo. The Hudsonian godwits I'm watching in a Rainwater Basin wetland have a life history that engages a third of the Earth's landmass and two of its oceans. Their annual migration, one of the most extreme of any animal on land, at sea, or in the air, requires flight biomechanics and neural capacities that would be wasted on the cardinals at my backyard bird feeder. Humpback whales and Hudsonian godwits, should they make eye contact, would give each other knowing nods. Yet here they are, taking a breather in Nebraska.

While shorebirds softened my heart toward the Rainwater Basin, it was a seabird that melted it altogether.

Franklin's gulls spend much of their lives at sea but head to the prairie marshes of interior North America to propagate the next generation. Their inland migration brings them through the central Great Plains in big numbers and adds another dimension of wonder to the Rainwater Basin wetlands.

Franklin's gulls have one of the longest migrations of any of the world's gulls. Wintering primarily along the Pacific coastline of South America, mostly from Peru through Chile, they may travel as much as five thousand miles to reach their northernmost breeding areas in central Canada. They begin arriving in the Rainwater Basin in late March, peak in numbers in the latter half of April, and have pretty much cleared out by the end of May. They come back through at the end of summer, mostly in late September and early October.

Gulls were my least favorite group of birds until I got my first close look at a flock of Franklin's, hunkered down on a Rainwater Basin wetland on a blustery day in early spring. These handsome creatures are smaller and more delicate in form than the gulls you see in coastal areas and, when they arrive in fresh breeding plumage, are striking with their dazzling white underparts, jet black head, and slate gray wings.

But what takes your breath away is the blush of pink that suffuses their bright white breast at this time of year, inspiring "Franklin's rosy gull" as an early common name. The color has been described as "peach-blossom-red" and reads a bit like a wash of flamingo pink, not surprising since, as in flamingos, the tint is due to a pigment consumed through a diet of marine crustaceans.

In addition to good looks, there is something captivating about the way a flock of these birds glides across the landscape, gracefully mirroring the undulations of the terrain and lifting up here and there en masse to clear a fence or power line. Words like "lilting," "buoyant," and "bouncy" have been used to describe the flight of Franklin's gulls, which Nebraska ornithologist Paul Johnsgard declared "extremely beautiful." The best time to catch the sight is when cold, windy conditions keep them flying low as they move between wetlands.

The common name for this gull comes from *Larus franklinii*, published in 1832 in *Fauna Boreali-Americana*, a treatise on the zoology of the northern parts of "British America." The name honors Sir John Franklin, British naval officer and Arctic explorer, who disappeared with his crew while trying to navigate the Northwest Passage. The description was based on a specimen collected in present-day Saskatchewan in 1827.

While "Franklin's gull" endures as the common name, ornithologists now recognize a different scientific name as having precedence. *Larus pipixcan* was published in 1831 based on a specimen taken somewhere in Mexico. Ornithologists eventually recognized that these were two names for the same species and, because it was published one year earlier, *Larus pipixcan* was given primacy. The scientific name is now *Leucophaeus pipixcan*, the genus name inspired by the "shining white" plumage.

We needed to wade through this bit of nomenclatural tedium to get at the marvelous origins of the word *pipixcan*. The name was originally recorded by Spanish naturalist Francisco Hernandez de Toledo in his *Historia Avium Novae Hispaniae*, published in Rome in 1651 but based on his expedition to the New World in the 1570s. Hernandez's name was taken up by German ornithologist Johann Georg Wagler, who published the name *Larus pipixcan* in 1831 based on that specimen from Mexico. This strange word, it turns out, was derived from an ancient Aztec name for "gull."

All of this naming business is testimony to the monumental efforts of the Old World to make sense of the biodiversity of the New, which in the case of Franklin's gull involved a cast of explorer-naturalists, sea captains, patrons of scientific societies, and kings and other assorted monarchs from Spain, Germany, and Great Britain. But what I find stirring is the thought of an Aztec watcher of birds bestowing a name on a creature that spends part of its life flouncing over Rainwater Basin wetlands.

Farmers on the plains have their own name for *Leucophaeus pipixcan*. Franklin's gulls are famous for following tractors during the spring planting season, foraging for worms, grubs, and other critters turned up by plowing, and in the process giving the driver a close-up view of the birds in their finest breeding plumage. Writing in 1891, Kansas ornithologist Nathaniel Stickney Goss noted that Franklin's gulls were "called by the farmers the Prairie Dove."

Of course, the bird is a gull and not a dove, but most nineteenth-century Kansas farmers had never been to the seashore and, in their experience, doves were the loveliest of birds. You can hear the affection in this earthy name.

For each of the four hundred-some species described in *Birds of Nebraska*, the authors provide "finding" tips—suggestions of regions and habitats where the reader is most likely to encounter that particular bird. Delightfully, the entry for Franklin's gull says that during peak migration periods the species "can often be found by simply looking into the sky."

I highly recommend trying this tactic on a sunny day in October. No, it doesn't always work and, yes, people will wonder what's wrong with you, but to look up and discover wind-tossed, Pacific-bound Franklin's gulls, their brilliant white undercarriage flashing against a cobalt-blue autumnal sky, is one of the great birding experiences of the country of the Kaw.

We can't leave the subject of a Rainwater Basin seabird without mention of the black tern. Black terns start arriving in the region toward the end of April and are passing through well into June. Headed to breeding areas in the marshes and wetlands of the northern United States and central Canada, these sleek birds spent the winter along the coasts of Central America and northern South America, or simply just hung out at sea. With black head and underparts and dark gray wings and tail, black terns look a bit like bats as they float over a Rainwater Basin wetland, dipping down to pluck insects from the water surface.

Franklin's gulls and black terns each bring a distinctive elegance and animation to the Rainwater Basin in spring, plus the delightful paradox of their dual citizenship, being birds of ocean *and* prairie. I admire them both, but it is the prairie dove that thrills me most, swirling back into the region by the tens of thousands, white breast blushed pink from a diet of Chilean seafood, the aura of the Aztec empire in its name.

The Rainwater Basin is not the only part of the country of the Kaw that supports migrating shorebirds and seabirds. Saline wetlands along the eastern edge of the Smoky Hills and playa lakes on the High Plains provide valuable stopover habitat for these birds. And recent surveys in the Kansas Flint Hills have found that upland pastures are important to American golden plovers, upland sandpipers, and buff-breasted sandpipers as they wing their way to and from the Pampas grasslands of South America, earning the region designation in 2016 as a Landscape of Hemispheric Importance by the Western Hemisphere Shorebird Reserve Network. But none of these areas support the seasonal concentrations of shorebirds found in the Rainwater Basin.

The life histories of close to fifty Western Hemisphere shorebird and seabird species intersect the Rainwater Basin, making the region an unlikely nexus between the birdlife of North and South America. But this is also some of the best corn country in the Cornhusker State. So, the nexus is a fragile one.

Ironically, it was hunting that provided the impetus for conservation efforts in the Rainwater Basin. Recognition of the region's importance to ducks and geese led the US Fish and Wildlife Service to begin acquiring property in the region in the 1960s. Today, this federal agency manages sixty-one Waterfowl Production Areas in the Rainwater Basin, and the state of Nebraska has established another thirty-five Wildlife Management Areas in the region.

Conservation in the region took a huge leap forward with the establishment of the Rainwater Basin Joint Venture in 1992. A remarkably broad coalition of landowners, government agencies, researchers, agriculture businesses and associations, and others, the organization is dedicated to acquisition, restoration, and management of the region's wetlands and surrounding upland habitats.

While these efforts were originally focused on waterfowl, what has been good for ducks and geese has also been good for dunlins and godwits. It has also benefited peregrine falcons, magnificent raptors that follow flocks of migrating shorebirds like gray wolves did the bison. In fact, your best chance of seeing a peregrine in Nebraska is in the Rainwater Basin in spring.

Around the time the Rainwater Basin Joint Venture was established, Joel Jorgenson and other ornithologists were beginning to discover and document the importance of the region to migrating shorebirds. Thanks to the research and advocacy of Joel, Drs. McCarty and Wolfenbarger of the University of Nebraska-Omaha, and others, the Rainwater Basin is now recognized by the state of Nebraska as a Biologically Unique Landscape and a priority for conservation efforts.

As for me, I can't imagine another year of life without spending time in the Rainwater Basin. I can't say that I'm completely free from my aversion to wetlands, but I've met the buff-breasted sandpiper, the Franklin's gull, and a host of other extraordinary trans-hemispheric travelers.

Awe trumps fear.

ELEVEN

The Waters of Mother Kaw

> Mother Kaw is on the move this morning...
> *Up a Creek Canoe & Kayak Rental, Facebook post on January 31, 2019*

The earliest image of the Kaw I've come across is a painting by Alfred Jacob Miller, a nineteenth-century American artist famous for his depictions of landscapes and characters associated with the Rocky Mountain fur trade. Miller composed his watercolor around 1858 but based it on field sketches he made in 1837 while part of an expedition that launched from the environs of today's Kansas City to the annual fur-trader's rendezvous in the Green River Valley of Wyoming. Titled *Crossing the Kansas*, the painting is in the collection of Miller's works at the Walters Art Museum in Baltimore. I have not seen it in person but I have seen its near-duplicate, *Forming Camp*, held by the Joslyn Art Museum in Omaha.

The painting depicts a large company of men, horses, wagons, and other equipment making its way from the south bank of the Kaw to the north. Historians believe the crossing took place at a ford about eight miles to the west of today's Lawrence. The ambience is luminous, with gauzy clouds in a pearlescent sky and the sheen of sunshine on the water. Two words come to mind: pristine and idyllic.

I've struggled with what to say about the Kaw as a river, that is, how to talk about the life within its waters. To begin with, I'm a botanist with very little knowledge of aquatic ecology. But that was something I could remedy by reaching out to fishery and stream biologists who study and care for the waters of the Kaw. My inquiries all seemed to loop back to Dr. Keith Gido at Kansas State University, who carries the title of distinguished professor and leads the university's Fish Ecology Lab. Keith and his students and colleagues have published extensively on the ecology and fishes of Great Plains streams, the Kansas River basin in particular.

But rather than pristine and idyllic, these research papers are often peppered with words like "degraded," "highly endangered," "alarming," even "bleak."

Like all rivers except those in the most remote and unpeopled parts of the Earth, the Kaw and its tributaries have been profoundly changed—channels reengineered, natural streamflow fragmented and disrupted by reservoirs and other manmade impediments, waters polluted by municipal sewage and agricultural operations or waylaid altogether by groundwater-mining irrigation systems.

Such changes to the aquatic environment have of course had tremendous impacts on aquatic life. I wondered if, like bison ecology, the aquatic ecology of the Kaw was something I would have to talk about in the past tense.

Thankfully, stretches of near-pristine waters still exist within the country of the Kaw, and these treasures provide windows into the finely tuned workings of healthy aquatic ecosystems. As O. J. Reichman wrote in *Konza Prairie*, such streams "are the bows on a landscape that is wrapped in prairie." But life endures even in the more altered parts of the Kaw and its tributaries, and that is a gift worth opening as well.

The main stem of the Kaw flows 173 water miles from the merging of the Republican and Smoky Hill Rivers to confluence with the Missouri. Within this reach the Kaw grows into a big river, in places running with a bank-to-bank (nonflood) top width of eight hundred feet or more. The big river is also a big presence, a felt presence, that inspires Lawrence resident Craig Pruett, owner of Up a Creek Canoe and Kayak Rental, to speak of his beloved river as "Mother Kaw."

Big rivers harbor big fish, and the lower reaches of the Kaw has produced some doozies, notably catfish (blue, channel, and flathead) but also paddlefish and sturgeon (pallid and shovelnose). These and other species were so abundant in the Kaw that a small community of commercial fishermen sprang up in and around Lawrence in the late 1800s, the men known locally as the Kansas Riverkings.

The earliest members of this unofficial fraternity were two black men, Jake Washington and Abe Burns. Jake arrived in the area sometime before 1870 and Abe a bit later. Their favorite spot was below the Bowersock Dam in what is now downtown Lawrence. A photograph from 1896 shows the men with a hoisted pair of blue catfish, one weighing 92 pounds and the other 110. The scene in the photo is replicated in a bronze sculpture that now stands in Abe and Jake's Landing, an event hall set along the river's edge not far from where the men had their fishing shack. Commercial fishing in the area continued into the 1960s, with characters like Gustave "Old Man River" Graeber and Maurice "Catfish" Wustefeld taking up the Kansas Riverking mantle.

During the glory days of the Riverkings, the Kaw was well endowed with sandbars, islands, shallow side channels, and quiet backwater areas. The result of periodic floods, these physical features of the riverscape were in a constant state of flux among genesis, obliteration, and recreation, giving the Kaw a diverse and dynamic habitat structure. Sandbars are still common in certain stretches today, providing great places for paddlers to picnic or camp as well as critical nesting habitat for piping plovers and least terns.

The Kaw and its tributaries have been subject to a number of floods in historical times, some of which have been monumental. The Great Flood of 1844 is relatively unknown because it occurred before there were towns, roads and other infrastructure, or even many farms in the Kaw watershed, but it exceeded in height and spread the floods of 1903 and 1951, which thoroughly inundated the industrialized bottomlands of Kansas City, Kansas. Then there was the 1935 flood in the Republican River basin that nearly washed away entire towns in Nebraska. The loss of life and property caused by these devastations prompted the creation of a series of reservoirs on upstream portions of the Kaw watershed to help control floodwaters and, in the drier western reaches, support agricultural irrigation.

Today, eighteen federal reservoirs, operated by the US Army Corps of Engineers and the Bureau of Reclamation, are scattered throughout the watershed of the Kaw. From west to east, the largest of these are Swanson, Harlan County, and Milford on the Republican, Cedar Bluffs and

Kanopolis on the Smoky Hill, Wilson on the Saline, Tuttle Creek on the Big Blue, Perry on the Delaware, and Clinton on the Wakarusa. The Kaw itself is impounded at Lawrence by Bowersock Dam, a low hydroelectric dam.

The ecological impacts of these water projects are many. For certain fish species, the rising water levels of spring provide an environmental cue that triggers spawning, and artificial stabilization of river levels by reservoirs disrupts this natural seasonality to the detriment of reproduction. For others, reservoirs and other man-made structures fragment stream connectivity, preventing movement of fishes above and below the structure. And lack of periodic scouring by high waters lessens the natural formation and maintenance of sand bars, backwaters, and other physical aspects of the riverscape that are required by different species, from the huge paddlefish to the diminutive plains minnow.

The Kaw was recently added to the Sustainable Rivers Program, a joint effort of The Nature Conservancy and the Army Corps of Engineers to find ways to manage dams and other water projects to optimize benefits for people and nature. Heidi Mehl of the Nature Conservancy, who leads this effort in Kansas, hopes that more natural flow regimens can be established for the Kaw while flood control and irrigation needs continue to be met.

The creation of this system of reservoirs resulted in another unintended ecological consequence for the Kaw. Sand from the mainstem of the river is of a size and shape that makes it desirable for manufacturing concrete, and a substantial commercial dredging industry has existed for decades along the lower reaches of the Kaw. But mining of sand and gravel creates deep holes and pockets in the riverbed that the river seeks to fill, to the harm of the riverscape.

Prior to completion of the upstream reservoirs, large quantities of sand were naturally transported to the mainstem of the Kaw by its network of tributaries. But with this transport impeded by reservoirs, the river is forced to take sand and silt from within its own environs; this results in bank erosion, channel widening, degradation of sand bars, and increased siltation of the river bottom, all of which alter the riverine habitat to the detriment of aquatic life.

In addition to alteration of the physical structure of the river environment, the waters of the Kaw have also suffered greatly from pollution, especially the stretch through the bottomlands of Kansas City, Kansas, which for years received unimaginable amounts of organic wastes and other toxins from meatpacking plants, refineries, and sundry heavy industries as well as residential sewage from a good part of the metropolitan area.

A federal investigation of water quality in the Missouri River basin in the late 1960s found "conditions of serious pollution" all along the lower reaches of the river, but especially as it passed through Kansas City. In one facet of the study, wire baskets containing live channel catfish were placed into the Missouri at select points along its course, including at confluence with its tributaries, to determine the effects of water quality on the taste of the flesh after four days of exposure.

Test fish placed into the Kaw scored near the bottom of the flavor acceptability rating system. Even worse, fish placed into the Blue River, just eight miles downstream, died within four hours. But the most frightening results came a bit farther downstream at the confluence with Sugar Creek, where researchers pulled up their basket and found only skin and bones—the flesh had been completely dissolved by substances in the water.

Thankfully, regulations established by the federal Clean Water Act of 1972 have resulted in much better conditions today, but water quality of the Kaw is still impacted by inputs from municipal wastewater facilities and industrial discharges, as well as fertilizers and pesticides washed into the river and its tributaries by agricultural run-off and urban stormwater run-off. Yet, while not pristine, the Kaw is now healthy enough to support recreational activities, and canoe and kayak rental companies like Craig Pruett's do business from Junction City to Kansas City.

The resiliency of the aquatic environment, aided by enforcement of water quality and dredging regulations, is evident on a morning when competitors in the All American Catfish Tournament set out from the Kaw Point boat ramp in Kansas City, Kansas. Blue and flathead catfish still get beefy in the Kaw, with fifty pounders not uncommon. Local resident Tim Berger used to manage this catch-and-release tournament, which draws professional anglers from all over the United States. Tim says the Kaw offers some of the best catfishing in the country, not only because of the size of the fish but also because the river is more manageable in a smaller boat than the Missouri.

While baskets of channel cats revealed environmental peril in the 1960s, the catch of contestants in the All American Catfish Tournament offers testimony that the waters of Mother Kaw are on the mend.

Compared to a hulking flathead catfish, the southern redbelly dace is a petite, flashy jewel of a thing. And in contrast to the murky big-river world of the flathead, this little minnow dwells in the bright, clear-running reaches of Flint Hills creeks.

We've already considered the importance of the Flint Hills to the preservation of tallgrass prairie, one of the most diminished and imperiled ecosystems on Earth. But the Flint Hills also provides critical refuge for a band of fishes that prosper in rocky, spring-fed watercourses surrounded by healthy rangeland. The list includes southern redbelly dace and a company of other sprightly daces, darters, and shiners, most of them under three inches in length.

The rocky and rugged landscape of the Flint Hills keeps most of the land cover in tallgrass prairie; it also favors the formation and persistence of springs. The ridge and slope topography of the region is controlled by a series of limestones that are harder and more resistant to erosion than the shales with which they are interbedded. Rainfall penetrates deeply into the cracks and fissures of the limestone, but when the accumulated water hits an impervious layer of shale, it travels along the surface of the formation until it can exit in a ravine or some other low point between hills. Streamflow is also supported by gravelly banks that absorb and store rainfall as it moves down into the channel from the uplands.

The streams that arise in these hills are among the most pristine in all of the Great Plains. Mill Creek is a shining example. Flowing into the south side of the Kaw about twenty miles west of Topeka, the upper watershed of Mill Creek is mostly cloaked with tallgrass prairie and gallery forest, and it supports native fish species that have disappeared from streams running through farm lands.

Kings Creek is a similarly healthy Flint Hills stream that arises on the Konza Prairie Biological Station, jointly owned by Kansas State University and The Nature Conservancy. The Kings Creek Basin is part of the Hydrologic Benchmark Network of the US Geological Survey and is the only stream in the network that exclusively drains near-pristine tallgrass prairie.

Much of the work of the Fish Ecology Lab at Kansas State University has been focused on Kings Creek, which is only a ten-mile drive from campus. Long-term monitoring of fish communities and habitat began in 1995 and is ongoing. In addition to Kings Creek, Keith Gido and his students, including Dr. Erika Martin (now on the faculty at Emporia State College), have a complementary long-term monitoring program in the Fox Creek watershed on the Tallgrass Prairie National Preserve, forty-five miles to the south near Strong City.

Many of the small fishes that inhabit these prairie streams sport flamboyant coloration during breeding season and have lively names to match. The most striking are southern redbelly dace, carmine shiner, Topeka shiner, and plains darter. These are often accompanied by the less-showy but still appealing bluntnose minnow, plains stoneroller, common shiner, creek chub, johnny darter, and, in a few places, blackside darter. These assemblages often include slender madtom and stonecat, two diminutive species of catfish that only occur in clear, quick-flowing streams with rocky bottoms.

None of these fish species is strictly limited in distribution to the Flint Hills. All of them range more widely, at least historically, to the east and north, particularly in the rocky streams of the Ozark Highlands. The endangered Topeka shiner was first collected in Shunganunga Creek near Topeka and appears to have been most common historically outside of the Flint Hills; it now finds refuge in the unmuddied waters of this rocky region.

I witnessed the beauty of these creatures up close when I attended the third annual Fish Wizards Tour in 2022, an educational event held on a tributary of Mill Creek south of the village of Alma. Sponsored by the Native Stone Scenic Byway organization and the Volland Foundation, a local art gallery and community gathering space, the event is led by knowledgeable and enthusiastic scientists who carefully seine fishes from the stream and share information on their biology with participants. It struck me that day that these fishes are to the aesthetics of prairie waters what warblers are to oak woodlands.

And, similar to the way warblers partition the resources of a woodland in spring, fishes in Flint Hills streams have fine-scale habitat preferences: riffles versus pools, shallow pools versus deeper pools, rubbly versus fine-textured substrates, open runs versus stretches with in-stream cover (rock ledges) or canopy cover (tree shading). Some prefer the headwaters of the creek, while others occur farther downstream. Because these preferences are most strongly expressed during the spawning season, having the right habitat available at the right time is critical to reproductive success and the ongoing welfare of these species.

As in every other ecological system, the greater the habitat diversity, the greater the fish diversity. But the dynamic nature of Flint Hills streams, with sudden flooding in spring and drying of headwaters in summer, makes watershed-scale connectivity an especially important goal of conservation efforts.

The majority of land in the Flint Hills is in private ownership, and most of the owners have a strong conservation ethic. This is certainly in evidence on Mill Creek and its tributaries, which are rightly famous for healthy waters and rich communities of prairie fishes. But given the fragile nature of tallgrass prairie streams, it is comforting that the watersheds of Kings Creek and Fox Creek are safeguarded within the bounds of protected nature preserves.

If life can be precarious for a minnow inhabiting the headwaters of a Flint Hills creek, consider the upper reaches of a High Plains stream. Annual precipitation in this semiarid environment is about half of what typically falls in the Flint Hills, and even that is negated by the drying effects of summer heat, high winds, and low relative humidity. And good years are often few and far between, interrupted by the droughts that are a regular feature of the High Plains climate. As a result, there is not enough input from the sky to maintain streamflow except after a cloudburst. That such streams carry water at all is by the grace of the Ogallala Aquifer.

Prior to the advent of irrigated agriculture, exposures of water-bearing Ogallala Formation sediments in gullies, ravines, and canyons of the High Plains supported innumerable springs and seeps. Writing of the fishes of upper Republican and Smoky Hill watersheds in 1940, biologist John Breukelman reported that where these streams cut their valleys into the Ogallala strata, "they are supplied by springs that insure almost a continuous flow even during long periods of drought." It is worth noting that his observations came after the devastating Dust Bowl drought of the 1930s.

As I contended earlier, streams birthed and supported by this groundwater were part of the rich fabric of natural resources that made the country of the Kaw, at least for a while, the greatest bison country in all of bison country. It also made it pretty fine minnow country.

We'll call them the Ogallala minnows: brassy minnow, plains stoneroller, creek chub, fathead minnow, suckermouth minnow, northern plains killifish, red shiner, sand shiner, and plains darter (not a true minnow). Each has a biology and life history that allows it to live at the extreme limits of aquatic life, where streams are ephemeral, carrying surface water only during or shortly after a storm, or intermittent, receiving enough groundwater seepage to maintain short segments or at least pools in most years. To inhabit this highly dynamic stream environment requires hardiness and also the capacity to retreat and recolonize as conditions warrant.

The ambassador for this group is the northern plains killifish, "a fish," wrote grassland ecologist David Costello, "I always greet as an old friend." It is hard not to like this little minnow with its bold, zebra-like striping. In addition to its delightful appearance, plains killifish prefer to school in very shallow water, seldom more than six inches deep, making them easy to spot from the bank. Then there is the fact that the plains killifish is a true creature of the Great Plains, occurring nowhere else—one of the very few fishes whose world is entirely limited to the grasslands.

The plains killifish is as tough as they come and is one of the last of the Ogallala minnows to drop out of the mix as stream conditions deteriorate. Although it was once ubiquitous in the Kaw's western tributaries, its distribution and relative abundance have declined significantly since the 1970s. Dr. Costello's little friend is tolerant of shallow water, warm water, alkaline water, and highly saline water, but it cannot endure the absence of water.

I was driving through a remote part of the Wray Dune field in southwestern Nebraska on a hot August day when I crested a hill and came upon a picturesque sweep of sandsage prairie. I stepped out of the car to take some photos but was immediately distracted by a sound that didn't jibe with the parched, silvery landscape before me. I followed the noise to the other side of the hill and discovered an expansive stretch of bright green corn, the hissing spray from the center pivot irrigation ricocheting loudly off the stiff, waxy leaves.

Sandsage prairie occurs in the driest parts of the western Great Plains, where in a good year the land receives eighteen inches of precipitation. A typical corn plant needs about twenty-six inches of water over its five-month growing season. You do the math. To grow corn in sandsage country you have to make up the difference. You have to tap the Ogallala.

It is no secret that the Ogallala Aquifer is in trouble. Depletion of the aquifer has been occurring since the 1960s, when advances in pumping and irrigation technology first made it possible to raise corn in sandsage prairie country.

The problem is, water in the aquifer is fossil water, accumulated over millions of years when mountain-born streams washed over what would later become the Great Plains, and there is little to no recharge of the aquifer in the semiarid climate of the High Plains. You can keep drilling deeper and deeper into the sediments of the Ogallala, but when the water is gone, it is gone.

Long before pumps tapping the Ogallala began to run dry, agricultural irrigation lowered the water table enough that the springs and seeps that supported High Plains streams stopped flowing and groundwater was no longer supplied to the streambed. In the language of hydrologists, these streams became "decoupled" from the aquifer—a clinical term for what to a fish is an ecological catastrophe.

The impact has been documented in studies throughout the watersheds of the Republican and Smoky Hill Rivers, where these unique High Plains fish communities have been greatly diminished or erased altogether by aquifer depletion.

It appears that the best hope in the country of the Kaw is the Arikaree River in eastern Colorado, especially an eight-mile stretch on Fox Ranch, a Nature Conservancy preserve in Yuma County that still supports a healthy community of Ogallala minnows, among them the brassy minnow, here at the very western limits of its range in the central Great Plains.

The socioeconomic implications of aquifer depletion are profound. The development of aquifer-supported irrigation on the High Plains gave rise to a regional agricultural economy that today is almost entirely dependent on groundwater mining, especially with the rise of industrial-scale, Big Ag enterprises like concentrated animal feeding operations (cattle feedlots, hog confinement barns) and ethanol plants. Depletion of the Ogallala not only threatens the existence of individual farms in the region but also has the potential to unravel the fabric of entire rural communities.

The big question is whether some kind of sustainable level of groundwater extraction can be achieved that would slow or even halt the demise of the aquifer. All kinds of groups are working on this: state and federal agencies, university researchers, nonprofits, agricultural lenders, commodity groups, seed companies, ag chemical companies, irrigation equipment manufacturers, even youth organizations, plus a host of coalitions with "Ogallala" in their names.

While the survival of farms, Main Street businesses, schools, churches, and other human institutions is at stake, here's hoping these efforts also benefit one of the most audacious biological communities in all of the country of the Kaw—the Ogallala minnows.

Among the materials Keith Gido sent me to read was *Big Creek and Its Fishes*, a forty-three-page booklet published in 1986 by Fort Hays State University. He described it as "obscure, but really insightful." I'm glad he brought it to my attention.

Big Creek is a tributary of the Smoky Hill River that arises on the High Plains of western Kansas near Grinnell and wanders eastward for 163 miles, often within eyesight of Interstate 70, connecting with the Smoky Hill just south of Russell. The military outpost of Fort Hays was established on Big Creek in 1867, and this stretch of the creek now flows through the town of Hays and the campus of Fort Hays State. The publication was the result of a couple of graduate students in the biology department wondering if anything actually lived in the "much-maligned creek." Surprisingly, their survey work turned up twenty-eight different species of fish.

The story of Big Creek is emblematic of almost all High Plains waters: its upper reaches now disconnected from groundwater due to mining of the Ogallala Aquifer, its lower reaches muddied by silt washing in from wheat and corn fields, and its native fish diversity diminished.

A glimpse of the past splendor of this little stream is provided in the poignant forward to *Big Creek and Its Fishes* written by Dr. Gerald Tomanek, then president of Fort Hays State. The creek flowed through the Gove County farm where Dr. Tomanek was born in 1921, not far from its headwaters, and served as the local swimming and fishing hole: "The water was crystal clear and there were pools twenty feet deep with a gravel or rock bottom."

While Big Creek may be a shadow of its former ecological self, it has two claims to fame in the annals of ichthyology. First, it is the type locality of the amiable northern plains killifish. Historical research by Mark Eberle, retired aquatic biologist from Fort Hays State, indicates the type specimen of the species—the specimen that served as the basis for the original scientific description and naming of *Fundulus kansae*—was netted around 1884 in Big Creek near the town of Ellis.

Second, it launched the artistic career of Joseph Tomelleri. Joe, then a graduate student, was the lead author of *Big Creek and Its Fishes* and also prepared the illustrations that accompanied the booklet. Since photographs rarely depict fish well enough to aid identification, and Joe had some artistic ability, he decided to take a stab at drawing the subjects of the booklet with colored pencils. He would eventually become the John James Audubon of fish artists; his thousand-some scientifically accurate illustrations have appeared in over five hundred places, from fishing magazines to outdoor clothing catalogs to field guides. The definitive *Kansas Fishes*, published in 2014 by the University Press of Kansas, contains 184 of Joe's beautiful full-color drawings. From Big Creek to the big time!

Thinking back to the idyllic stream of his boyhood days, Gerald Tomanek pined, "It would be nice to find a way to rejuvenate Big Creek back to the way it was sixty years ago." That is a pretty tall order.

As the university president noted at the time, some streams in the eastern United States had been restored to health, but those renewals were achieved primarily by reducing water pollution within the watershed. Certainly, that has been the case in the more urbanized stretches of the Kaw, as evidenced by the success of the All American Catfish Tournament in waters that were once all but unlivable.

Not to dismiss such conservation victories, but the situation of

Big Creek and other Great Plains streams is far more complicated and challenging. Restoring the ecological health of these waters is not just a matter of reducing pollution, it requires protecting or, in most cases, re-establishing connection to the groundwater on which their very existence depends. To be brutally honest, even if there was the resolve to manage the Ogallala Aquifer for the benefit of aquatic life as well as human welfare, in most parts of the plains it is just too late.

This harsh reality makes the preservation and protection of streams like Mill Creek, Kings Creek, and the Arikaree River of utmost importance. There are no more priceless and vulnerable natural treasures in all of the country of the Kaw.

TWELVE

Beautiful Contrivances

> All in all, there is hardly to be found for effective
> cross-fertilization a more wonderful coordination of
> complex structures in flowers with the structures and
> ways of insects.
> *William Chase Stevens,* Kansas Wild Flowers

Spiderworts have a way of making the lovesick more sick. In his 1929 book, *Prairie Smoke*, ethnobotanist Melvin Gilmore recounts a bit of Lakota folklore about a young man wandering alone on the prairie, lost in thoughts of his sweetheart, when he encounters a spiderwort plant in bloom. In the beauty of the flowers he sees the girl's beauty, and the overwrought lad breaks out in a love song—*to the plant*.

What is so stirring about spiderworts? Maybe it is the color of their flowers, some of the truest blues to be seen on the prairie. Then there is the timing—they bloom in spring when young men and women are at their mooniest. But there is a loveliness to the form and frame of these grasslike plants that unsettles even mature botanists. Some of the imagery they use to portray spiderworts—"graceful lines"; "delicate texture"; "rare and elegant coloring"; "tender-bodied"; "dewy"—could have been lifted from the Song of Solomon, and leaves you feeling like you need to take a cold shower.

While we humans invest flowers with all kinds of emotion and meaning, to a plant these structures are just the body parts by which it makes the next crop of seed. Sexual reproduction, whether in plants or animals, requires the union of a male sex cell (sperm) with a female sex cell (egg) to produce an embryo. Since plants can't pursue their mates, they rely on pollen grains to carry the male sex cells to the abode of the female sex cells. Wind is the pollen vector for grasses and most trees, but a whole host of our wildflowers enlist the aid of animals, primarily insects, which they attract by dangling the reward of pollen or nectar within their charming flowers.

The nuts and bolts of pollination involve the transfer of pollen grains from a stamen to a pistil. Stamens are the male part of the flower and are essentially a filament capped with an anther in which the pollen grains are formed. The pistil is the female part and is composed of an ovary that contains eggs awaiting fertilization, topped by a narrow style topped by a stigma, the tip that receives the pollen. The pistil is located in the center of the flower, with the stamens generally arrayed to the outside. After successful pollination and fertilization, the pistil matures and ripens into the seed-bearing fruit.

Beyond this basic template, plants display fantastic and sometimes bizarre variations in the shape, size, and arrangement of floral structures, often tied to the anatomy and biology of a specific type of insect. Charles Darwin referred to these as "beautiful contrivances." A multitude of these amazing plant/animal interactions can be witnessed in the country of the Kaw.

All of our early-flowering woodland herbs are pollinated by insects, which is remarkable since this is a dicey time of year for a cold-blooded bug. There is always the chance of a fitful start to spring that throws off the timing of insect emergence and foraging activity. Yet this is the time when the maximum amount of sunlight reaches the forest floor to run the photosynthetic engines of the plants. So somewhere along the line the plants and insects in these communities worked through the cost-risk-benefit analysis, and we get the wonder of a woodland spring.

Studies of the flowering ecology of spring woodland herbs have shown that native bees and flower flies are the major floral visitors and potential pollinators. There are many intriguing relationships between these plants and their insects, but none more amazing than Dutchman's breeches and bumblebees.

Dutchman's breeches are so named because of the shape of its dangling flowers, which look like little white hearts or, as someone fancied, upside-down pantaloons. The spurs that form the baggy pant legs bear nectar glands that must be accessed by pushing through the opening of the flower—the tight "waist" of the breeches—where the pollen-bearing anthers are held close by two inner petals.

In the Midwest, flowers of Dutchman's breeches are pollinated almost exclusively by two-spotted bumblebees, which have the size and strength to push into the flower and reach the nectar. In the process, the bee picks up pollen on its body and carries it off, to be deposited when it enters another flower. So, the whimsical form of the flower, a delight to our eyes, is the outward expression of an internal structure closely tied to the shape and dimensions of a bumblebee's head.

But the bond runs even deeper. The only bumblebees available when Dutchman's breeches start to bloom are young queens recently emerged from their solitary, belowground dens after hibernating through the winter. Impregnated at the end of the previous summer, the queens rumble about the woodland in search of nectar to replenish depleted fat reserves so they can get about the monumental work of constructing a nest, laying eggs, and raising the next generation of bumblebees.

Based on meticulous observations, Ohio botanist Lazarus Walter Macior concluded that the flowering of Dutchman's breeches is actually synchronized with the emergence of queen bumblebees from hibernation. It seems that there are all kinds of things that could go wrong in such an entangled relationship at such a capricious time of year, resulting in either fewer seeds or bumblebee larvae, or both, but these creatures are fully committed to the partnership.

If I had to choose a wildflower to serve as the floral emblem for the prairies of the Kaw, it would be purple coneflower. There are nine different species in the genus *Echinacea*, and three occur in our region—narrow-leaf, pale, and Topeka purple coneflowers. Each has a relatively long season of bloom, from May into July.

These iconic plants are members of the Sunflower family, and what looks like a single flower is actually a composite structure called a head, with tiny disc florets in the center and ray florets circling the outer edge. The disc florets, two to three hundred of them, are packed onto a raised, cone-shaped structure called the receptacle. Each ray floret produces a single, strap-shaped ligule, rose pink to reddish purple in color, which droops down like a petal. The overall effect is like a clunky rocket in flight, with vapor trails streaming behind.

Narrow-leaf purple coneflower, the most widespread of our three species, occurs throughout the shortgrass and mixed-grass prairie regions and in tallgrass prairie in the northern part of the Flint Hills. The smallest of the three, it is generally about two feet tall but is often shorter on dry, rocky sites. Pale purple coneflower is native to tallgrass prairies mostly east of the Flint Hills and may reach three feet in height. It is taller and more willowy, and the ligules of its ray florets are about twice as long as those of its western relative, up to three inches, which gives the flower heads a spidery appearance and makes them much more animated in the wind. Topeka purple coneflower has very short ligules, about an inch long; instead of drooping, they are strangely curved inward. This rare plant has a very limited range, occurring in scattered upland sites from the central Flint Hills south into Oklahoma and Texas.

A wide variety of butterflies and native bees work over the tube-shaped disc florets of purple coneflowers for nectar or pollen or both. The most distinguished of these visitors is the regal fritillary butterfly. With a wingspan of three to four inches and striking coloration—bright red orange and blue black, checkered with white markings—regal fritillaries are a thrill to behold as they cruise above the tallgrass canopy. Butterfly researcher Ann Swengel called the regal fritillary "prairie royalty."

Although regal fritillaries visit a variety of wildflowers, the abundance of purple coneflowers in the forb mix makes the plant a frequent nectar source, and the domed central cone of the flower head provides a prominent elevation for displaying the butterfly's splendor. Regal fritillary caterpillars, however, are exceedingly picky when it comes to flora and feed only on the foliage of violets. In our area they feed almost exclusively on prairie violet, a grassland specialist.

Adult females lay their eggs in leaf litter at the end of summer, and the larvae, which hatch in the fall and are dormant through the winter, commence feeding on violets the following spring. Diminutive, gemlike plants, prairie violets are an ephemeral presence in the visible flora; they send up new growth in March, flower in April, then dry up and go dormant for the rest of the growing season.

Regal fritillary larvae prefer young, fresh violet leaves, and, after reaching full development, they enter the pupal stage. By the time regal fritillaries emerge from this stage and are on the wing, the violets are overtopped by the new growth of prairie grasses. The entire reproductive strategy of this eye-catching butterfly is molded by the all-but-invisible violet.

To watch a regal fritillary nectaring on a purple coneflower is to witness the essence of the tallgrass prairie. In 2020, the US Mint issued an America the Beautiful Quarter for the state of Kansas that depicts a regal fritillary floating above sprigs of big bluestem and Indiangrass. The Mint called the butterfly "iconic to the Tallgrass Prairie National Preserve." I wish the artist had slipped a few purple coneflower stems into the composition.

That such a scene is a fairly ordinary occurrence in our tallgrass prairies belies the fact that regal fritillaries have "suffered extraordinary decline" across their formerly expansive range. These spectacular butterflies were once relatively common throughout the central and eastern United States, from prairies of the interior to meadows on the Atlantic Coast. Thankfully, regals have a stronghold in the country of the Kaw, particularly between Kansas City and the Flint Hills. A survey in 2005 of eighty-seven tallgrass prairie remnants in northeastern Kansas revealed a regal fritillary metapopulation of "tremendous size and conservation significance."

The sighting of regal fritillaries brings joy, not only because of their beauty but also because their presence is an affirmation of ecological health in the tallgrass prairie. Perpetuating that joy requires a vigorous population of prairie violets, and that requires careful use of prescribed fire to maintain microhabitat conducive to violet flourishing. To use Aldo Leopold's imagery, purple coneflowers, regal fritillaries, and prairie violets are cogs and wheels of the prairie mechanism, and fire is the tinkering needed to keep it all running.

I owe a lot to a little bitty milkweed.

In the world of milkweeds, the dwarf milkweed is a pretty big deal.

Growing less than three inches tall (the species name *uncialis* means "inch-high"), it is the smallest of the nearly thirty species of milkweeds (genus *Asclepias*) that inhabit the Great Plains, some of which can reach six feet. It is also the first to bloom, opening clusters of rose-purple flowers as early as the end of April. And to top it off, it is one of the rarest milkweeds in the United States.

I undertook a study of the dwarf milkweed because it had been identified as a plant of conservation concern for the Great Plains—perhaps in danger of extinction. My main task was to look for this plant in places where it had been found in the past, to see if it was still there. Most of the historical occurrences were on the eastern plains of Colorado and New Mexico where shortgrass prairie is the dominant vegetation.

I succeeded in finding a few plants in a handful of historical locations, including a site on a grassy slope above the South Fork of the Republican River in eastern Colorado. But most of the twenty-some places I searched had none, even when the habitat was intact and in fairly good condition. It appears the dwarf milkweed was more common in the past than it is today.

My relationship with the dwarf milkweed yielded unanticipated benefits. First was the glory of becoming the world's leading expert on this plant. I had never before been the world's leading expert on anything, so I welcomed the honor, even if the circle of admirers was as tiny as the plant itself.

Pursuit of the dwarf milkweed also gave me a deep dive into a part of the country of the Kaw that was barely known to me. The painstaking search for this diminutive plant required walking a lot of shortgrass prairie, and walking it very slowly. In the process I gained an intimate familiarity with this grassland that I never would have had otherwise.

Finally, getting to know this little plant opened my eyes to the flabbergasting milkweed way of making more milkweeds.

About 130 species of milkweed are native to North America. The country of the Kaw has twenty-one, from the purple milkweed of Kansas City–area woodlands to the dwarf milkweed of the shortgrass plains. Our flora also includes the beloved butterfly milkweed, with its vivid orange-red flowers, and Mead's milkweed, an endangered tallgrass prairie specialist that has become extinct throughout much of its former range but is hanging on in some of our eastern prairie remnants.

The structure of the milkweed flower is one of the marvels of the plant world, matched in complexity only by that of orchids. While floral anatomy differs greatly between milkweeds and orchids, both uniquely package their pollen within waxy, sac-like structures called pollinia.

Pollinia occur in pairs in an arrangement that looks something like a saddlebag, with two sacs full of pollen grains dangling from thread-like straps that are connected to a central knob called a corpusculum. A milkweed flower has five of these contraptions, called a pollinarium, each associated with a vertical slit within a five-angled boxlike structure at the center of the flower called a gynostegium, which encloses the female floral parts. The corpusculum is situated at the top of the slit, while the pollen sacs are tucked away just inside the chamber.

For most insect-pollinated plants, the mechanics involve dusting some part of the bug's body, typically the mouthparts, head, or thorax, with lots of individual pollen grains as they rummage around in the flower seeking nectar or pollen. The milkweed pollen delivery system involves packets of pollen and engages the legs and feet of insects as they crawl across a flower cluster sipping nectar.

The process begins with a rambling insect stepping into a milkweed flower. As it withdraws its leg, wing-like ridges on the sides of the central chamber guide it upward along a slit toward a projecting corpusculum. If a hair on the leg should snag this knob, and the insect is strong enough, it will pull the entire pollinarium out of the chamber as it extracts its leg.

With the pair of pollen sacs attached to its appendage, the insect continues to mosey across other flowers in the cluster or flies off to work over a different cluster. Pollen transfer is accomplished when the insect steps into a second flower and, its leg once again guided upward by the ridges on the chamber, slips the pollen sac of the first flower into a slit of the second flower. As the insect removes its leg, the sac breaks loose and—voila!—a package of five hundred pollen grains has been delivered.

It gets even crazier. The pollen sacs dangling from the corpusculum have to be reoriented before they will fit into the slit of the second flower and deliver the goods. This is accomplished when the pollinarium is yanked from the first flower and the straps holding the sacs, called translator arms, dry in the air while the insect is en route to the second flower and twist the requisite ninety degrees.

This madcap scheme depends on big-bodied bugs strong enough to pull themselves free of the slit. Bumblebees and other large bees, as well as wasps and bigger butterflies, are up to the task, but smaller, weaker insects may get trapped and have to leave a leg behind in order to escape. It also depends on a high volume of insect foot traffic, so milkweed flowers produce lots of nectar and are typically highly fragrant. These enticements seem to work since blooming milkweeds are usually swarming with six-legged visitors.

I recall being deep in the Nebraska Sandhills one day, soaking up the silence, when I detected a pervasive but low-frequency hum in my ears. It turned out that sand milkweed was all around in the prairie, and the sprawling plants, in full bloom, thronged with blissfully buzzing bees, wasps, and flies.

The genus *Asclepias* made it into *The Origin of Species*, the first edition published in 1859. Darwin was fascinated by cases of coadaptation between plants and animals, and he found it remarkable that the "very curious contrivance" of the pollinarium occurred in both orchids and milkweeds, "genera almost as remote as possible amongst the flowering plants." He discussed the matter, along with the structure of the eye, under the heading of "Organs of extreme perfection and complication" in a chapter dealing with what he saw as "difficulties" with his theory of descent by modification.

Just over a century later, Kansas botanist William Chase Stevens was likewise moved by the milkweed flower. Stevens described and illustrated the workings of dozens of plant–pollinator relationships in his delightful book *Kansas Wild Flowers*, but he was most fascinated by the "wonderful coordination" between milkweed flowers and milkweed pollinators.

At the twenty-third annual meeting of the Kansas Academy of Science in 1890, B. B. Smyth gave a paper titled "Periodicity in Plants." In it he offered observations of "daily motions in plants," specifically, what time of day their flowers opened for business and how long individual flowers remained open and viable. He summarized his meticulous observations in "A Floral Clock for Kansas," a creative compilation of fifty-two plants, mostly Kansas prairie wildflowers, arranged by time of flower opening, from hedge bindweed at 2:00 a.m. through small-flowered gaura at 8:00 p.m.

One of the last species on the list is Missouri evening primrose, clocking in at 7:00 p.m. on Smyth's botanical timepiece. This showy wildflower occurs in rocky prairie uplands in the Flint Hills and is widely grown in gardens for its large, moon-yellow flowers. The flowers are saucer shaped, about four inches across, and they taper abruptly into a long, slender floral tube up to five inches long. The lower part of the tube holds a reservoir of nectar.

The blooming season of Missouri evening primrose begins mid-May and extends well into June. An individual flower opens just around sunset and wilts at sunrise, although it may last a while longer if the sky is overcast.

The flowers are pollinated almost exclusively by hawk moths, which resemble hummingbirds in size and shape and manner of flight. Like the sunset-opening flowers, hawk moths are vespertine—of the evening—and take to the wing at dusk. And, corresponding to the floral tube, the tongue of the moth can be amazingly long, as much as four inches in species that visit Missouri evening primrose.

In the mechanics of this relationship, the hawk moth hovers over the sky-offered flower, unfurls its coiled tongue, and inserts it into the center of the shallow saucer and down the floral tube to reach the nectar reward. In the process the moth brushes against the flower's anthers and is dusted with pollen grains. The moth then disengages by flying backward (an aerial maneuver otherwise limited to hummingbirds and dragonflies) and moves on to another flower, plunges in for a sip, and in the process transfers pollen from the first flower to the stigma of the second.

A study in the Flint Hills found that six different species of hawk moths visited Missouri evening primrose flowers; the white-lined sphinx was the most abundant and reliable pollinator throughout the flowering season. This guild of hawk moths appears to have worked out a temporal partitioning of the rich floral resources of the evening primrose: three of the species are active in the early part of the flowering season and then drop out of the mix, and two other species arrive on the scene a bit later.

While the twilight ballet of hawk moths and Missouri evening primrose flowers is a thrill to behold, it overshadows the exquisite relationship between this wildflower and the evening primrose sweat bee. This small bee is a member of a small genus of bees that only use pollen from evening primroses to provision the nest for their young, and they have bodies and behaviors tightly bound to the peculiarities of the evening primrose flower.

First is the unique way evening primrose pollen is packaged, with the pollen grains strung along strands of sticky, cobweb-like threads. The female evening primrose bee has a uniquely modified apparatus called a "scopa" on its hind legs that enables it to efficiently comb up and carry clumps of pollen-packed threads.

Then there is the challenge of foraging at dusk and dawn when the flowers are open but the light is low. Bees, the vast majority of which are day-flying creatures, see by means of a pair of large multi-lensed compound eyes on the sides of their heads and small single-lensed ocelli ("simple eyes") on top of their heads that aid in navigation and orientation. The evening primrose sweat bee and its "night wandering" relatives have much larger ocelli than typical day-flying bees; the lenses gather in enough light, even moonlight, to guide the bee on its rounds.

A bit of entomological history makes this tale even more wonderful: the evening primrose sweat bee was first collected in the Flint Hills, captured from the flowers of Missouri evening primrose plants growing near Blue Rapids, Kansas, in 1919.

The spiky rosettes of yuccas are a familiar sight throughout the central and western parts of the country of the Kaw. While the bayonet-like leaves are intimidating, getting close enough to take in the beauty of a yucca flower is worth the risk of getting poked.

Yuccas start to bloom toward the end of May. The plants send up a stout flower stalk two to five feet tall that displays a couple dozen nodding, bell-shaped flowers, two inches long and creamy greenish white in color. Cradle one of the lily-like flowers in your hand and carefully turn it up toward you. If it is getting near dusk, you're likely to see yucca moths spilling out of it.

Most of our wildflowers, whether woodland or prairie, have working relationships with a variety of insects, and most of these insects likewise forage from a fairly eclectic floral menu. Such insects are generalist pollinators, and the relationship between plant and pollinator, while mutually beneficial, is not exclusive. But the yucca plant and the yucca moth have put all their eggs in one basket and have an obligate mutualism. Neither species can propagate the next generation without the other.

Yucca moths are about an inch long and silvery white in color. They emerge in spring from underground cocoons and head to freshly opened yucca flowers to find mates, drawn by fragrance emitted by the flowers at twilight. Once impregnated, females get after the work of rearing the next generation. They live only three to five days, so there is no time to lose.

The first step involves gathering pollen from anthers within a yucca flower and packing the sticky stuff into a ball about the size of her head. The female then tucks the ball under her chin and flies off to a different flower; she enters and seeks out a place on the seed-producing ovary to pierce and lay an egg. That done, she heads to the top of the pistil and stuffs a small amount of pollen from the ball into an opening, forcing it down inside using mouthpart appendages unique to yucca moths. She may repeat the process in the same flower or fly to another.

The larvae that hatch from these eggs feed on yucca seeds developing within the ovary; they consume many but not all. After reaching a final stage of development, the larvae burrow out of the seed capsule and, descending to the ground on a slender thread, work their way a couple of inches below the surface and spin a cocoon. Meanwhile, seeds that escaped the larvae are ripening and, if all goes well, the ovary will mature into a dry, woody pod with enough seeds inside to rattle in the wind.

Just when you think you have plumbed the depths of some biological wonder, you discover there is more. A yucca plant has the capacity to sense when the numbers of moth eggs laid in a particular flower's ovary are high enough to compromise its capacity to form seed, and will abort that flower to prevent further investment of resources. But abortion of the flower means that moth eggs will be lost as well. A female yucca moth is wired to know this and will limit the number of eggs inserted into a yucca flower by depositing a pheromone during egg laying, a chemical signal that marks the flower and informs the next visitor that she was there first, inducing the second female to either lay fewer eggs or just move on to another flower.

Producing a good crop of seed is only part of the business of reproduction. To be truly successful, the seed needs to be dispersed to favorable habitat where it can germinate and grow into a new plant. Many plants on the plains rely on wind to do the job. Nuttall's violet has worked out an arrangement with ants.

Nuttall's violet, a diminutive wildflower of the shortgrass prairie, blooms and sets seed in early spring amid clumps of blue grama and buffalograss. Attached to the surface of each seed is an elaiosome, a fleshy structure rich in lipids and proteins. Ants gather the seeds and carry them off to their nests, where they either eat the elaiosome or feed it to their larvae, discarding the actual seed as so much refuse, within the nest or scattered about outside.

So, the ant gets a meal and provision for its young, and the little plant hunkered down in the shortgrass turf gets its seeds hauled away. This interaction is called myrmecochory, roughly "dancing with ants," and is an example of mutualism, a relationship that benefits both ant and plant.

In a final dazzling act of collusion, the time at which the violet's drying seed capsules split open and release seed is synched with the time of day (between 9:00 a.m. and 1:00 p.m.) when ants are most active but mice and other rodents, who enjoy elaiosomes but also consume the seed, are tucked into their hidey-holes, snoozing away.

As if such intimate intertwining of plant and animal being isn't enough to make your head spin, bear in mind that the astonishing elaboration of flower structures and pollination strategies demonstrated across the world of plants, these "beautiful contrivances," contribute absolutely nothing to the survival of a particular plant in a particular place at a particular time. It is all about providing a successful launch to the next generation. The virtuosity of life is often most extravagantly displayed in the perpetuation of life.

And consider that the whole enterprise is star driven. Take a Missouri evening primrose plant growing on a rocky ridge in the Flint Hills. The plant will begin to bloom on a day in the month of May when Earth reaches a particular point on its annual half-billion-mile journey around the sun. The opening of an individual flower will be triggered when Earth's daily rotation causes the Flint Hills to slip away for a few hours from the face of the sun, a signal that also sets hawk moths and evening primrose bees to wing. Our evening primrose plant, riding a circuit of the universe on planet Earth, is keeping an eye on our star.

THIRTEEN

The Hard Places

> The autumn light was growing dull about me, the shadows were gathering. I was beyond the country of common belief; that would seem to be the source of my problem. I had spent a lifetime exploring questions for which I no longer pretended to have answers.
> *Loren Eiseley,* All the Strange Hours: The Excavation of a Life

You cannot be a bona fide nature writer from Nebraska and not at some point quote Loren Eiseley. Born and raised in Lincoln, Eiseley graduated from the University of Nebraska in 1933 then pursued his interest in anthropology at the University of Pennsylvania, receiving his PhD in 1937. He landed his first teaching appointment at the University of Kansas, where he was on the faculty until 1944.

Although an academic anthropologist by profession, Eiseley is best known as an essayist and poet who pondered deep questions raised by what he saw in Nature. I find a lot of his writing to be dark and grave, but I do admire his honest recognition of mystery in the natural world and the limits of the human mind to come to grips with it.

One of my favorite Eiseley quotes comes from his first and most widely read book, *The Immense Journey*. Among life's "strangest qualities," he wrote, is its "eternal dissatisfaction with what is, its persistent habit of reaching out into new environments and, by degrees, adapting itself to the most fantastic circumstances."

Eiseley was musing about life in the deepest depths of the ocean, but the same could be said of the country of the Kaw, where remarkable creatures inhabit the most impossible of environments. The surprising biota associated with Ogallala rimrock and High Plains playas testifies to this resolve of living things to find some kind of place to do their living, to fill every nook and cranny, even the hard places.

If you've taken a summertime hike above timberline in the Rockies, you may have noticed that the most charming wildflowers occur in the harshest of habitats—fellfields. *Fell* is Gaelic for rock, and these areas are literally pastures of stone. Sun-burned and wind-blasted, a fellfield is the most xeric environment in the alpine, yet is populated by the most captivating of plants. Compact and low-growing, most just a couple of inches tall, these cushions, carpets, and tufts are smothered in flowers, making mountain fellfields look like miniature gardens.

Similar plant communities occur in the country of the Kaw in association with exposures of Ogallala Formation bedrock. The setting may be a rimrock ledge above a canyon or the gravelly crest of a butte. The soil, such as it is, will be shallow, minimally developed, and stony, with little moisture-retaining organic matter. Coupled with full exposure to the sun and wind, this is the most severe plant habitat in the Great Plains.

Ogallala fellfields have a fairly consistent guild of what we'll call rock plants. The most common as well as most striking is silky milkvetch, which grows as a prostrate mat up to three feet across but barely one inch tall and, when in bloom, is blanketed with tiny rose-purple, pea-type flowers. It is frequently accompanied by stemless four-nerve daisy and depressed nailwort, which spread into less expansive mats, the cushion-forming Hooker's sandwort, and more open-growing but tufted species like lavender-leaf evening primrose and plains phlox. Silver-mounded cryptantha, oval-leaf bladderpod, and summer milkvetch are additional cushion and mound formers that add to the visual drama wherever they show up.

While these species represent a diversity of plant families, they share a number of characteristics that enable them to flourish in a very difficult environment. First is stature. These low-growing plants have highly modified stems that are very short and condensed with little space between leaf-bearing nodes. They emerge at ground level from a thick woody base called a caudex, so it is hard to tell where stems begin and rootstock ends. Repeated branching of the caudex belowground gives rise to the low, circular pattern of growth characteristic of cushion plants, while the wide-spreading mat of plants like silky milkvetch forms by repeatedly branching stems that radiate out from the crown. These growth patterns keep rock plants out of the worst of the wind, which desiccates tender leaf and flower tissues in an already dry environment.

Another characteristic is reduction and simplification. The leaves and flowers of rock plants are often smaller in size and simplified in structure compared to their relatives from less challenging environs. Silky milkvetch provides an example. Unlike most species in the genus *Astragalus*, which have leaves comprised of twenty to thirty leaflets, the tiny leaves of silky milkvetch are divided into only three leaflets.

The name "silky" speaks to another common trait. The foliage of many rock plants is densely covered with tiny hairs called trichomes, giving a gray or silvery cast to the plant. The exposed habitat of a rock outcrop is subject to both intense direct sunlight and reflected light from the surface of the rocks; protective hairy foliage diffuses light before it reaches the sensitive tissues beneath and insulates against water loss and windborne grit. In most rock plant species the trichomes are just simple hairs, but those of the bladderpods, genus *Physaria*, are astonishingly elaborate and, under magnification, resemble microscopic starfish and snowflakes.

While rock plants are tough as nails, they wouldn't last a day out in the surrounding shortgrass prairie, where blue grama and buffalograss not only carpet the ground but also fill the upper soil horizon with dense systems of fibrous roots. Still, in a rocky environment where small fissures or cracks may be the only place suitable for root growth, the advantage goes to the taprooted cushion and mat plants. As inhospitable as it is, rock outcrop habitat offers escape from the competition of grasses.

The distinctive growth forms displayed by silky milkvetch and Hooker's sandwort are adaptations for survival at the limits of botanical life. You can find plants of nearly identical stature and form flourishing in other extreme habitats around the world, like cliffs overlooking the Arctic Ocean or the alpine zone in the Himalayas. That they grace such hostile and seldom-if-ever-seen places with such beauty seems extravagant beyond words.

You can view Ogallala rock gardens on public lands like Enders Reservoir State Recreation Area in Nebraska and Lake Scott State Park in Kansas, or anywhere else in the headwaters region of the Kaw where a road intersects rocky hills and bluffs. These communities are at their most glorious in late spring and early summer, mid-May to mid-June, when the majority of the rock plants bloom. But even when peak flowering is past, tapestries of interwoven silvery-gray cushions and mats reward a look-see. And who knows, you might even run into a rock wren.

Rock wrens are an unexpected joy of the Ogallala rimrock. The charm of a rock wren is not in its appearance but in its personality. The small, pale-gray things are hard to spot among the rocks, and you can easily overlook them until a male decides to break into song. In typical wren fashion, the singing is delivered with a volume and intensity that seems all out of proportion for such a diminutive creature. An individual male has an amazing repertoire of fifty to 130 song types, a play list from which they appear to select certain numbers depending on the need of the moment. The rocky habitat serves to amplify their exuberance.

Follow the din and you'll soon locate the performer, skittering across the ledge then stopping for a moment to size you up, bobbing up and down in agitation, only to plunge over the brink and down the bare rock face. Writing of his experiences with rock wrens in western Nebraska, biologist John Janovy Jr. rightly called this animated little bird "a bundle of fire."

Rock wrens are widely distributed throughout the western United States, from the Pacific Coast through the Rocky Mountains and onto the plains, and they occur in all manner of arid, stony places from desert mesas to mountain slopes to seaside cliffs. In the country of the Kaw, which represents the far eastern edge of the species' breeding range, rock wrens can be found in breaks and canyons of the High Plains and chalk badlands in the Smoky Hills.

These feisty little birds are wonderfully adapted to life on the open rimrock. Rock wrens are nimble and camouflaged; and, as far as anyone knows, they never need a drink of water, apparently getting enough fluid from a diet of insects and spiders gleaned from cracks and crevices.

But the most remarkable adaptation involves their nest. Rock wrens seek out crevices in an outcrop or spaces between boulders in which to place their nests. Tucked away inside the cavity, the nest itself is a soft cup made of sticks, stems, bark, grasses, feathers, and fur that rests on a foundation of small stones. This bed, which provides a level, somewhat elevated surface for the nest, is covered with a "pavement" of smaller stones that extends just outside the nest cavity entrance like a runway.

The stones that make up the pavement have been gathered and carried one by one to the nest cavity. The female of the breeding pair does most if not all of the hauling and, even in situations where the male does pitch in, appears to be the chief architect. An observer in New Mexico watched a female rock wren carry stones to a nest cavity at an average rate of one stone per minute for over two hours, while the male of the pair "perched on a nearby rock and sang." The wrens select stones that are mostly flat and relatively thin, apparently because they have better stacking and wedging properties compared to round stones.

Construction of a wren patio involves an enormous energy investment by these little birds, as revealed by the research of University of Northern Colorado biologist Lauryn Benedict and her student Nate Warning. Nate *counted* the number of stones used in forty-six inactive nest cavities; while the average number was 222, several nests contained more than five hundred stones. He also weighed the stones from each nest site and found that in some cases over two pounds had been hauled in, which is about sixty-two times the average body weight of a rock wren. That would be like the much more fit me of my high school years gathering up and toting home nearly five tons of rocks.

There has been a lot of speculation over the years as to the purpose of the pavement. Nate's research indicates the stones serve to decrease the size of the opening to the nest cavity; larger openings contain significantly more stones. A tighter opening may deter certain predators, primarily snakes and small mammals, and help reduce flooding of the nest cavity from intense rainstorms. It may also keep nestlings from falling out.

The pavement may also serve as something of an alarm system. Observers have long noted that the stacked layers of stone make a bit of noise when touched. Writing about rock wrens in his 1909 treatise, *The Birds of Washington*, William Leon Dawson remarked the stones "rattle pleasantly every time the bird goes in or out." The sound and/or vibrations of stone clinking against stone may signal the nesting female that her mate has just arrived at the entrance with a bit of food. It may also alert her of the approach of a predator, giving enough advance warning for the vulnerable bird to exit the cavity and either defend her brood or escape with her life.

While this behavior is intriguing, it is also extraordinary. Rock wrens are the only bird species in North America to make extensive use of rock in and around their nest site. But even more impressive is that they deliberately select stones of particular dimensions for their pavements and will vary the number used and their placement depending on the unique dimensions of the nest site cavity. This demonstrates a high level of behavioral flexibility and has the hallmarks of tool use, an exceedingly rare phenomenon in the animal world.

It is one thing to live some of your life in an extreme environment; it is another to raise a family there. Rock wrens have made the Ogallala rimrock home, in the deepest sense of the word.

While an exposure of bedrock is a very severe environment, it is a relatively stable one. Should a silky milkvetch seed find a place to germinate and the seedling maintain a foothold, the plant can enjoy an austere but generally settled life for many years to come. Studies of alpine plants have found that many attain a ripe old age—estimated at over three hundred years for moss campion, a dominant wildflower in North American alpine plant communities. Judging from the size and spread of mats and cushions growing on an Ogallala rock outcrop, it seems likely that our High Plains rock plants are long-lived as well.

Unstable habitats are far more challenging to living things because they require the capacity to adapt to the dynamics of a fluctuating environment. There is no more unstable habitat in the country of the Kaw than a playa lake.

We've already met Hal Borland, the ten-year-old picking flowers for his mom out on the shortgrass prairie. Borland wrote two books about growing up on the plains of eastern Colorado in the early 1900s, *High, Wide, and Lonesome* and *Country Editor's Boy*. The former contains a delightful description of a playa, which he and his father encountered as they rode in a wagon across the shortgrass prairie:

We came to a broad, shallow lake, melt from the winter snow and drain from the spring rain, that would shrink to a little mudhole in another month or two. It hardly looked like water, it was so clear, and under the water the grass was growing, much greener than the flat all around it. Ducks were there, scores of teal and mallards and even canvasbacks, swimming on the clear water over the green grass as though they were swimming in the air.

Borland experienced the playa in May, at the height of its glory, the shallow basin filled out to its margins. In addition to ducks swimming over buffalograss, he saw curlew, killdeer, and other shorebirds working the edges of the playa. As the wagon drew closer the ducks took off, dripping so much water "that for a moment there was a flash of rainbow in the spray." I've never come across a more wonderful portrayal of a playa in pristine condition.

Playas are ephemeral wetlands and, as Borland noted, can transition from inundation to dust in a relatively short period of time. While *ephemeral* is a pretty word, the environmental reality of ephemeral habitat is far from kind. Space and time are very limited in temporary habitats like a playa, requiring flexibility and sometimes spectacular, even aggressive adaptations to make a go of it.

Playas support a fairly consistent and unique assemblage of invertebrates. Among the most common are the branchiopods—clam shrimp, fairy shrimp, and tadpole shrimp. These macroinvertebrates are freshwater crustaceans that feed on algae, bacteria, smaller invertebrates, and general detritus. Largely absent from permanent waterbodies, branchiopods are considered signature species of the playa fauna. They are often most abundant, even dominant, in early stages of inundation (the hydroperiod), which is a boon to migrating shorebirds and other playa visitors.

Branchiopods deal with the transitionary nature of playa habitat by producing two types of eggs—thin-shelled summer eggs that hatch immediately and thick-shelled "resting" eggs that can lie dormant in the dried mud for extended periods of time and hatch when conditions are favorable. Resting eggs are highly resistant to cold, heat, and desiccation, and can even be carried on the wind to other playa basins.

In a semiarid environment with short-lived lakes and crazy winds, what could make more sense than flying prairie shrimp?

If any toad or frog could be called cute, it is the spadefoot. With their big, round eyes, spadefoots have a kitten-like innocence about them. But at certain stages in their life cycle they are anything but adorable.

Often called spadefoot toads, these amphibians are actually classified as frogs by herpetologists. There are seven species of spadefoots in North America, all adapted to arid environments. The name refers to a hard nob on their hind feet that enables them to dig burrows for shelter. The plains spadefoot occurs in the country of the Kaw, where it is an important cog in the ecology of playas.

Plains spadefoots spend most of the day belowground and emerge after dark in search of insects and other small arthropods. In times of drought they can stay belowground for extended periods by reducing their metabolism and descending into a prolonged dormancy.

Spadefoots only head to water to breed, which, given the temporary nature of a playa, must be done in synchrony with rains that fill the basin. Spadefoot breeding is triggered by the sound and vibrations that accompany a hard summer shower—rain striking the ground above them and thunder. Once aroused, they emerge from their burrows and make for the nearest body of water.

Compared to other amphibians, spadefoot mating and egg-laying is done in hyperdrive. A playa full of amorous spadefoots, called a "breeding congress," is noisy and chaotic, and mating and egg-laying may be completed in only two or three days. It is said that the racket from a spadefoot-packed playa can be heard from two miles away.

The eggs of amphibians hatch into an aquatic larval stage that undergoes metamorphosis to eventually become an adult. Toads and frogs in the larval stage are what we call tadpoles or pollywogs. Spadefoot eggs hatch in two or three days; the tadpoles develop quickly and complete metamorphosis in thirty-six to forty days. By comparison, the larval period of leopard frogs inhabiting a languid farm pond near Kansas City may last up to one hundred days.

Not to be outdone by the likes of shrimp, spadefoot toads also use a two-pronged approach to perpetuating the species in the tumultuous playa habitat. Spadefoot tadpoles, omnivorous when they hatch from their eggs, feed on detritus and algae and occasionally grab a fairy shrimp. But if shrimp are abundant and the tadpole ingests some minimum number of them, it will develop into a full-on carnivore.

In contrast to the omnivores, carnivore tadpoles are larger and have a more massive head with enlarged jaw muscles. They also have shorter intestines, as befits a meat eater. And there are dramatic differences in behavior between the two morphs: omnivores graze in groups along the margins of the playa while carnivores cruise the open pool like lone wolves, searching for prey.

What does shrimp abundance have to do with the switch to a carnivore morph? As mentioned, fairy shrimp and other branchiopods are most abundant in the early stages of inundation, when the playa basin is reaching maximum fullness. But the days of mud and dust are just around the corner, and life in the playa is about to descend into ecological dystopia. Rising shrimp abundance appears to serve as an environmental cue to spadefoot tadpoles, a warning that they better speed things up. Carnivore tadpoles develop more rapidly than omnivores, which gives them a better chance of making it to adulthood in a rapidly drying playa. The tadpoles are accessing playa longevity indirectly by evaluating shrimp density.

Transformation of spadefoot tadpoles from omnivore to carnivore is a stunning adaptation to what ecologists have come to recognize as "community disassembly." When a playa basin starts to fill from snowmelt and rainfall, a community of plants and animals begins to assemble. But as the playa nears the end of its hydroperiod, decreasing habitat size leads to increased competition for resources plus more frequent encounters between prey and predator. Organisms either abandon ship by emigration, enter some kind of dormant stage and wait for better times, die from lack of water, or get eaten. The playa community *disassembles*.

In the spadefoot economy, this is where things really get ugly. Carnivorous spadefoot tadpoles not only chow down on fairy shrimp but also prey on other creatures, including their own mild-mannered omnivore kin. They are not just carnivores, they have become cannibals.

After recounting examples of viciousness in the world of bugs, Annie Dillard concludes, "Fish gotta swim and bird gotta fly; insects, it seems, gotta do one horrible thing after another." There's a lot that happens in Nature that I wish I didn't know about. Cannibalistic tadpoles, for one. This is information I would just as soon purge from my mind. Such behavior is one reason I chose botany as my path through Nature—plants just don't stoop to the lows you find on the animal side of the ledger.

I would like to tell you that I've found a way to reconcile the beauty and brutality of Nature, but I'm still working on it. Like Loren Eiseley, the more I've seen the less I can explain what I've seen. I've stared into the abyss of the unsearchable . . . and shuddered.

Each of us edges up to the brink of this chasm clinging to a different metaphysical rope. "I am an evolutionist," wrote Eiseley; yet, toward the end of his life, he courageously declared, "I had come to feel at last that the human version of evolutionary events was perhaps too simplistic for belief." I'm trying to be brave enough to face up to the deficits in my own worldview.

But just because you can't explain everything doesn't mean you shouldn't ponder the big questions. Here's one for you: why do living things even bother with places like Ogallala rimrock and High Plains playas?

PART III

Nature & Culture

> ... where is the writer with the gusto, the wild
> immeasurable power, the wide horizons, the sense of the
> universal in man, and his infinite variety, that the Plains
> might have produced, should have produced?
> *Mari Sandoz,* Love Song to the Plains

In her heartfelt ode to the Great Plains, Mari Sandoz laments the passing of a golden era of literary foment at the University of Nebraska that peaked in the 1890s and gave rise to the likes of Dorothy Canfield Fisher and Willa Cather. If Mari is right, and the plains provides an exceptional place to plumb what is universal in the human condition, then we should be able to unearth some such things in the country of the Kaw. What follows is a collection of soundings in the realm of environmental history, stories of human interaction with the natural world. In the country of the Kaw, as in every other place on Earth inhabited by our kind, nature has shaped culture and culture has shaped nature.

FOURTEEN

Paradise Undone

> As we drove into this beautiful spot I proclaimed, "Boys, we have got into paradise at last!"
> *James R. Mead,* The Saline River Country in 1859

It was February 1860 and James Mead was leading a small hunting party across the divide between the Smoky Hill and Saline Rivers. One day at sunset they climbed a rise and witnessed a scene that stopped them in their tracks. Before them stretched a rolling prairie landscape with uplands covered with bison, draws full of elk, deer, and turkey, a stream abounding in beaver and otter, and gangs of wolves trotting around. Mead would later call this country "God's great park."

The men got to work, setting up a base camp in the area that would eventually become Mead's hunting ranch. The stream and valley became known as Mead's Paradise, a sentiment that echoes yet today in the name of Paradise Creek and the town of Paradise with its signature post rock water tower. But the arrival of Mead and his band of hunters marked the beginning of a very violent end for this grassy Eden.

The country of the Kaw has been the setting for human-wrought environmental devastations of staggering magnitude. As painful as it is to face up to these, it is necessary to truly understand the land as we experience it today, and to provide some hope of better stewardship in the future.

William Blackmore was a British financier and knowledgeable collector of Native American cultural artifacts who traveled and hunted the plains in the 1860s and 1870s. Richard Irving Dodge dedicated his *Hunting Grounds of the Great West* to Blackmore ("keen sportsman, genial companion, firm friend") and asked the respected Englishman to write an introduction to the book. Blackmore took to the task with relish and began his comments with a tirade about the "wholesale and wanton destruction . . . of the buffalo." As evidence, he cited an experience in the country of the Kaw:

> In the autumn of 1868, whilst crossing the Plains on the Kansas Pacific Railroad—for a distance of upwards of 120 miles, between Ellsworth and Sheridan, we passed through an almost unbroken herd of buffalo. The Plains were blackened with them, and more than once the train had to stop to allow unusually large herds to pass. A few years afterwards, when travelling over the same line of railroad, it was a rare sight to see a few herds of from ten to twenty buffalo.

"Like some tremendous, crashing sound that ended abruptly just at the moment we turned to listen," is how environmental historian Dan Flores described the jarring suddenness of the bison's extermination from the Great Plains. It was a crash that reverberated throughout the heart of the North American continent and was the opening salvo of what Flores called "the largest wholesale destruction of animal life discoverable in modern history."

James Mead and his entourage were early contributors to the crash. These men fancied themselves hunters and sportsmen, but were really just opportunistic prospectors, trying to mine a bit of profit from an enormous amount of raw material. Mari Sandoz and other writers of Great Plains history called them "hide men" or "hiders," because the hide was what they harvested with their Sharps and Springfield rifles. The rest of the bison was essentially organic rubble.

Dodge's book includes an illustration titled "Slaughter of Buffalo on the Kansas Pacific Railroad," a dramatic drawing by Ernest Griset that became an iconic depiction of the plight of the bison. The image was reproduced in William Hornaday's 1889 work, *Extermination of the American Bison*, a widely read publication of the United States National Museum credited with stirring public support to conserve the last survivors of this vanishing creature. It is possible that this distressing scene from the country of the Kaw actually helped save the bison from extinction.

The destruction of millions upon millions of bison is an undeniably dark chapter in American environmental history. Less well-known but no less appalling is an enterprise that developed as a side hustle to the buffalo hide trade. Gray wolves, the alpha predators of bison, followed the herds so closely the big canines were known as the "buffalo wolf." The large-scale slaughter of bison provided easy pickings for the wolves, which were happy to clean up carcasses left as gifts on the prairie.

But the wolves' scavenging provided hide men with an unprecedented opportunity to add another line to their portfolios at a time when beaver was becoming scarce and wolf-fur coats were becoming fashionable. Market forces that stretched all the way to Europe brought cataclysm into the world of the buffalo wolf.

Mead gives a description of the practice of "wolfing" in his memoir *Hunting and Trading on the Great Plains, 1859–1875*. The first step was to shoot down two or three old bull bison in separate places near the camp and let the carcasses lie one night to attract wolves. The next night, just before dusk, the men would scatter poisoned bait in and around the carcasses, each bait containing a dose of strychnine, a neurotoxin. Mead understandably does not describe the suffering of the animal, but it was terrible; the ingested poison caused violent muscular convulsions and eventually death through asphyxiation. The next day the men would go in search of dead wolves and skin them to harvest the pelt.

Just as eyewitnesses could only guess at the numbers of bison roaming the plains during their heyday, there is no way of knowing how many wolves were traveling along with them. But Josiah Gregg, who observed wildlife along the Santa Fe Trail from 1831 to 1840, wrote of wolves, "There were immense numbers of them upon the Prairies."

The spectacular abundance of bison in the country of the Kaw apparently translated into wolf wealth. An article in a Topeka newspaper in 1859 spoke of hunting parties "starting out from our city on trips to the buffalo *and wolf grounds*" [italics mine]. After their first night of baiting bison carcasses on Paradise Creek, Mead and his men found eighty-two dead wolves. They would eventually kill "some 5000 of them."

As horrific as it was, wolfing led to even greater wildlife devastations.

During the winter of 1861–1862, Robert Peck and two companions set up a wolfing camp on Walnut Creek in the southwest part of the Smoky Hills. In addition to wolves, their baits killed numbers of coyotes and "little yellow foxes"—the beautiful swift fox of the High Plains. The wolfers were killing and harvesting animals they could not even put a name to.

Great numbers of ravens followed the bison herds, scavenging the remains of animals brought down by wolves and, later, by the hide men. The actual species was the common raven. Mead admired the big birds, calling them "much handsomer than the ordinary crow, larger and with a different voice, far more musical, full of antics and fun." Like the bison herds they shadowed, flocks of common ravens are now a thing of the past on the plains; the species retreated into mountains and deserts to the west and boreal forests to the north. It was not a peaceful passing.

Mead wrote that ravens would feast on the flesh of "the hundreds of wolves which we had skinned and left lying around over the prairie," and, as they ate the viscera, they would also consume the partially digested baits. The poison would in time kill the birds, and the prairie "would soon be dotted with the glossy, shining bodies of the defunct ravens, with an occasional bald eagle among them." It must have been a hellscape—the skinned and mutilated carcasses of bison and wolves scattered across the landscape, punctuated by the bodies of ravens and eagles.

Paradise undone.

The northbound migration routes of a number of shorebirds flow primarily through the Great Plains. Long-distance migrants like American golden-plover, buff-breasted sandpiper, and Hudsonian godwit fly the entire length of the western hemisphere, twice a year. Maps depicting their overall migration patterns often have an hourglass shape to them. Nebraska's Rainwater Basin is situated at the pinch of the hourglass, and provides crucial stopover habitat for resting and refueling.

But such a concentrated staging area presents a worrisome ecological bottleneck for the birds that depend on it. The perils of the pinch are tragically demonstrated in the story of the Eskimo curlew, a pigeon-sized shorebird that is believed to be extinct.

From historical accounts, the Rainwater Basin once hosted enormous flocks of these birds in spring as they made their way from the southern tip of South America to breeding grounds on the coast of the Arctic Ocean (hence the *Eskimo*!). In a 1915 article titled "The Eskimo Curlew and Its Disappearance," Nebraska ornithologist Myron Swenk recounted reports of these birds alighting in the Rainwater Basin and covering "forty to fifty acres of ground." That is about thirty football fields, teeming with curlews.

But the massive concentrations of this highly gregarious and relatively tame bird made them easy targets for hunters. In the 1870s, market hunters from Omaha would head out to the Rainwater Basin and shoot wagonloads of Eskimo curlews. Sometimes the birds were so numerous that, if the men had enough ammo, they would dump the wagons out and refill them with fresh birds, leaving piles of curlews "as large as a couple of tons of coal" to rot on the prairie.

The Rainwater Basin was not the only danger zone for Eskimo curlews. These shorebirds were slaughtered wherever they congregated during their annual migration; Swenk noted that in New England and the Atlantic states "profligate gunners ... poured leaden death into southbound flocks of these unfortunate birds." But the University of Nebraska professor and native son pulled no punches in implicating the citizens of his state in the demise of the Eskimo curlew: "In this deadly work the people of Nebraska ... to our lasting discredit played a conspicuous and all too effective part."

Occasional reports of possible sightings stir hope that the Eskimo curlew may still exist in some corner of its former global haunts, but there have been no scientifically accepted records for North America since 1962. Sadly, a bird with the word *Eskimo* in its name will never again spend a few days with us here in Nebraska.

There are many reasons to lament the extinction of a species, but the shattering sadness of this story is the loss of the Eskimo curlew mind. Imagine the maps these birds carried in their heads! Charts formulated through the trips of untold generations of their kind winging from the bottom to the top of the Earth and back again—sketches of Tierra del Fuego, Galveston Island, Franklin Bay, the Gulf of Saint Lawrence, and a host of other landscapes, including our humble Rainwater Basin. I wonder if there was a flash of light, a surge of joy, when the contours of the terrain below matched up with an image on file in the curlew mind.

The accumulated intelligence that allowed the Eskimo curlew to occupy its unparalleled niche on our unparalleled planet was a work of exquisite beauty and inscrutable complexity. Now erased, like a masterpiece destroyed by vandals.

Back in my high school days there was a concert venue in Kansas City called Cowtown Ballroom. It had a short run, 1971 to 1974, but brought some great bands to town.

A cow town (to be proper, a *cattle town*) was a frontier settlement that sprang up at the junction of a railroad and a cattle trail. The country of the Kaw had two legendary cow towns—Abilene and Ellsworth. Both were born along the line of the Kansas Pacific Railroad as it worked its way up the valley of the Smoky Hill River and received longhorn cattle driven north from Texas along the old Chisolm Trail. The cattle were loaded onto railcars for transport to eastern markets, primarily Chicago, where they would be slaughtered and processed. Cow towns, famously wild, even dangerous places, attracted gamblers, grifters, prostitutes, outlaws, bounty hunters, and all the other characters you expect to find in a classic Western movie.

Kansas City still likes to trade on the cow town mystique, but, truth be told, the Kansas City cattle business was basically a factory operation. From the stock pens to the killing floor to the knifemen, splitters, headers, rumpers, and other workers who dismembered the carcass, it was all about assembly-line efficiency. As historian John Herron noted, "Stockyard workers in Kansas City labored in an industry that was just as mechanized as steel and just as dangerous as coal." And, like these industries, it generated massive amounts of pollution.

It all began in 1870, when stock pens were developed near the confluence of the Kaw with the Missouri, as a place where cattle trains from Abilene and Ellsworth could be unloaded and the livestock fed and watered before they were sent on through a web of diverging railroads to eastern packing plants. But investors, primarily railroad men, saw an opportunity to develop a packing industry in Kansas City as well, with much of the initial capital coming from Boston financiers. The setting for this lucrative venture? The bottomlands of the Kaw.

In *Kansas City and How It Grew, 1822–2011*, University of Kansas geographer James Shortridge highlights the importance of these bottomlands to the history and economic development of Kansas City. He names eight separate "bottoms" in the metro area—"graceful crescents of land that have formed inside the bends of the Kansas and Missouri rivers." Four are associated with the Kaw: the West Bottoms on the south side of the confluence with the Missouri River, and, a bit farther upstream, the Rosedale, Armourdale, and Argentine bottoms.

As the population and economy of Kansas City grew, railroad companies began to lay tracks in the bottomlands, which set the stage for the rise of the meatpacking industry. The Kansas City stockyards began in the West Bottoms, but the packinghouses, rail yards, and other facilities associated with the meat industry soon sprawled throughout the adjacent bottoms. Eventually, all the big dogs from Chicago built plants in the bottoms—Armour, Cudahay, Swift, and others—and meatpacking became Kansas City's first million-dollar-a-year industry.

The ugly reality of the meatpacking industry is pictured in, of all places, the Missouri State Capitol in Jefferson City. In the House Lounge of the capitol building is the famously controversial mural *A Social History of the State of Missouri*, painted by Thomas Hart Benton in 1935–1936. The forty-foot-long mural portrays historical moments from across the state. In the part of the composition devoted to Kansas City, Benton included imagery from the stockyards.

The scene jarringly juxtaposes a black slaughterhouse worker, blood-spattered, set to bring a sledgehammer down on the head of a cow, next to what looks like a businessmen's luncheon. A speaker is addressing the civic meeting from a podium on the elevated stage, and seated just behind him on the dais is Tom Pendergast, infamous political boss and kingmaker of Kansas City. A line of scantily clad dancing girls at the other side of the genteel gathering hints at Pendergast's vice operations, which he ran from the unsavory environs of the West Bottoms.

As environmental historian Amahia Mallea describes in her book *A River in the City of Fountains*, racial and social disparities in Kansas City made it easy to consign both pollution and vice to the bottoms. The mostly poor minority and immigrant populations that took up residence in the "river wards" suffered the most from the environmental and social degradation, but they needed the jobs provided by the polluting industries and lacked the political clout to fight for better living conditions.

The primary source of pollution in the bottoms was the processing of thousands and then millions of livestock annually. A typical cow produces sixty pounds of manure every day, and the banks of the Kaw became sites for stockyard manure dumps. Chunks of by-product from the slaughter and packing operations were tossed into the river as well. Later, petrochemicals and other toxins from refineries and heavy industry enriched the soup, along with an inflow of residential sewage from the growing human population of the Kansas City area. It is hard to imagine that any living thing other than the nastiest of bacteria could have survived in the waters of the Kaw.

The saga of Turkey Creek is part of the sordid story. The last stream to enter the Kaw before its confluence with the Missouri, Turkey Creek arises in the rolling hills of suburban Johnson County, Kansas, and flows northeast into the Rosedale Bottoms. The lower course of Turkey Creek, now greatly altered, looped east for a short distance across the state line into Missouri, where it received the waters of O.K. Creek, and crossed back into Kansas for about a hundred yards before emptying into the south side of the Kaw.

The upper part of the Turkey Creek watershed includes picturesque stretches of rocky streambed with occasional low waterfalls and wooded areas like those found today in Shawnee Mission Park. Prior to being rerouted in the 1920s, the lower course was essentially an open sewer, especially since O.K. Creek drained a good part of the fast-growing Missouri side of Kansas City. Things were so bad that O.K. Creek and the final stretch of Turkey Creek were buried in huge belowground sewer pipes. This made for more pleasant air quality aboveground, but Turkey Creek still delivered an unceasing stream of Missouri and Kansas sewage to the Kaw.

Not surprisingly, the lower part of Turkey Creek also provided habitat for some of the most notorious criminal operations in the decidedly disreputable bottomlands of the Kaw, including a district known as "Toad-a-Loop." Initially settled by people who could not afford to live anywhere else, proximity to the Kansas-Missouri state line made Toad-a-Loop a bit of a no-man's-land in terms of law enforcement. Enterprising ne'er-do-wells took advantage of this situation and built a saloon astride the state line, with gambling operations located on the Missouri side of the establishment and a network of secret passageways that allowed patrons to pass unseen between the two legal jurisdictions.

Stockyard and meatpacking operations in the bottoms peaked in September 1918, when Kansas City processed a world-record 55,000 head of cattle in a single day and 475,000 for the month. The last packing plant closed in 1991, but the decline began decades earlier. In the opinion of historian Daniel Serda, the catastrophic Kaw flood of July 1951, at the time the costliest in US history, was the beginning of the end of the meatpacking industry in Kansas City. It destroyed much of the already-aging industrial infrastructure in the bottoms. It also devastated homes, schools, churches, and shops in the factory towns of Rosedale, Argentine, and Armourdale, forcing some twenty thousand residents to relocate, most of whom would never again live in the bottoms.

As a celebrated artist, Thomas Hart Benton traveled in rarefied social circles in his home town of Kansas City. But he also had a powerful affinity for working-class people. Moved by the immense suffering caused by the 1951 flood, Benton composed the painting *Flood Disaster (Homecoming—Kaw Valley)*, a heartrending scene of a family returning to their ruined home and belongings. Benton had lithographs made of the original painting (which sold in 2011 for nearly $1.9 million) with the title "Homecoming—Kaw Valley 1951" and sent them to every member of Congress to encourage government funding for disaster relief.

While the flood was indeed a catastrophe, there was a brutal silver lining—it precipitated the healing of what had become an environmental wasteland.

The dawn of the 1920s was a heady time for wheat farmers on the Great Plains. Demand for their crop was strong as Europe recovered from the devastation of World War I, and the price per bushel was climbing. Then there were those incredible new gasoline-powered machines rolling out of factories back East—tractors for plowing and cultivating and combines for harvesting and threshing—making it possible to farm more acres with less labor. Seizing on this favorable state of affairs, farmers pushed westward, farther out onto the High Plains, planting even more wheat on even bigger swaths of land.

The money was so good that a new breed of agricultural entrepreneur arrived on the scene—the suitcase farmer. Then as now, the type of wheat grown in the central and southern plains was hard red winter wheat, which has a unique cultivation cycle: seed is planted in September or October, the young sprouts grow a bit through the fall, overwinter as a clump of foliage, and then commence growth in the spring, producing heads of grain that can be harvested in early summer. The life cycle of the crop, coupled with the power and mobility of the new machinery and an improving network of roads, made it possible to plant wheat on a distant piece of ground in late summer, return home, and not come back until June or July the next year when it was time to harvest.

Often the absentee owner was a relatively small operator who lived and farmed a county or two to the east and was just trying to expand his holdings a bit, usually with the help of a bank loan. But many suitcase farmers were speculators who owned the land and hired others to plant and harvest: small-time investors like teachers, clergy, auto mechanics, and doctors, and those who had more resources like banks, implement dealers, oil companies, even insurance companies. The deeper the pockets of the speculator the bigger the scale of their operation and the greater the acreage of shortgrass prairie put to the plow. Most of these investors likely never laid eyes on the land itself.

The crop, the technology, the flowing cash, and the flat open country of the High Plains made possible the creation of what historian Donald Worster called "a vast wheat factory . . . a landscape tailored to the industrial age."

Then the Dust Bowl happened.

The Dust Bowl drought of the 1930s was the worst prolonged environmental disaster in American history, an ecological and socioeconomic catastrophe. The first really dry year in the Great Plains happened in 1931, but drought conditions worsened from 1935 through 1937 and persisted into 1940, making it all but impossible to raise a decent crop of anything on the plains for several years. Economic hardship caused by the drought was magnified by the Great Depression of 1929 to 1939, when millions of Americans were out of work and enduring financial hardship.

The feverish expansion of wheat acreage in the 1910s and 1920s resulted in the destruction of millions of acres of shortgrass prairie in the central and southern Great Plains. But crop failures coupled with plummeting commodity prices caused most wheat farmers to give up and just quit planting, and the recently "broken" ground was largely abandoned.

Land once cloaked with a sun-and-wind-forged turf of blue grama and buffalograss lay bare and open to massive erosion.

The most terrifying manifestation of these apocalyptic times, and what gave the region its name, were the dust storms. Sometimes the storm would arrive as a "black blizzard," a gigantic, rolling wall of dust, thousands of feet high, complete with thunder and lightning, "like a tornado turned on its side." At other times it would start more slowly but build to a howling crescendo that could last for days. No matter how one of these tempests began, the aftermath was a landscape choked in dust and sand.

The Dust Bowl was both an event and a place. When the geographic extent is mapped, the impacted region looks something like a slightly tilted oval that stretches through the plains from southwest Nebraska deep into the Texas Panhandle. The middle and upper reaches of the Kaw watershed fill the northern third of the oval, but the severest impacts were felt to the southwest where the far western corners of Kansas, Oklahoma, and Texas meet the eastern ones of Colorado and New Mexico. Certainly, the most iconic and tragic images of Dust Bowl days came from the southern plains, although there is a photo in the National Archives of a dust storm descending on the Republican River valley town of Naponee, Nebraska, on March 26, 1935.

Timothy Egan's *The Worst Hard Time* chronicles the stories of men, women, and children who went through the darkest days of the Dust Bowl. One of them was Don Hartwell, a farmer who tried to ride things out near Inavale, Nebraska, a few miles west of Willa Cather's Red Cloud. He and his wife, Vera, eventually lost their farm in 1940 through foreclosure to the bank, with the added indignity of it being auctioned off at a sheriff's sale. Despair is evident in the entry in his diary for November 8, 1936: "I burned some Russian Thistles on the W. place today. I cut down a dead tree W. of our house. I set this tree more than 20 years ago, it was a Norway poplar & it seemed that when it turned green that spring had really come. But the drouth of the last few years got it—the same as it has us."

What Dust Bowl scholars like Donald Worster surmised through historical research, climatologists have in recent years affirmed through sophisticated computer modeling. Recurrent periods of drought are a common feature of Great Plains weather, triggered by the interaction of sea surface temperatures with atmospheric circulation patterns. Yet the Dust Bowl drought of the 1930s was highly unusual for North America, and computer modeling of ocean–atmosphere inputs alone cannot account for the intensity of the dryness, the extreme temperatures, and the fearsome dust storms. But when climatologists add land-cover changes to their simulations, the makings of a perfect storm come together. Reduction in the temperature-buffering cover of vegetation, coupled with incalculable tons of soil particles swept up into the atmosphere, actually amplified what would otherwise have been a serious but fairly run-of-the-mill drought.

The Dust Bowl was born in the Great Plow-Up. Wheat, one of humanity's most ancient staples, had been weaponized.

In his essay "A Native Hill," Wendell Berry recounts the story of a company of men opening a road through the unbroken forest of eastern Kentucky in 1797. On the approach of one very cold night they felled great numbers of hickory trees and built huge bonfires. Then, for amusement, the men decided to divide into two teams, take burning logs from the fires, and throw them at each other. They engaged in this sport for a couple of hours until, with the fires burning out and a number of the men wounded, the match was stopped, the fires rekindled, and everyone lay down to sleep.

Berry shares this story because of its utter violence, from the orderly violence of clearing forest to the disorderly violence of the battle of the firebrands. Being a native Kentuckian, he admits to a bit of pride in the wild exuberance of his forebears. But Berry laments the tendency to environmental violence that seems inherent in our kind. His explanation is that the roadbuilders were "*placeless* people," adding, "Having left Europe far behind, they had not yet in any meaningful sense arrived in America, not yet having *devoted* themselves to any part of it in a way that would produce the intricate knowledge of it necessary to live in it without destroying it."

Looking back on the environmental violence inflicted on the country of the Kaw over the past 150 years, the same principle has been at work:

the worst devastations came and still come at the hands of people with no fidelity to the land. Thankfully, there are those who have *devoted* themselves one way or another to some part of the country of the Kaw. We'll meet a handful of them in the final chapter.

FIFTEEN

Community Ecologies

> ... as there are no little people in God's sight, so there are
> no little places.
> *Francis Schaeffer,* No Little People

Los Angeles–based journalist Corie Brown spent several weeks during 2016 driving the back roads of Kansas. A native of Wichita and a fourth-generation Kansan on both sides of her family, she returned to her home state to investigate the connection between the decline of rural communities and access to healthy food. She titled her piece, published online in The New Food Economy, "Rural Kansas Is Dying: I Drove 1,800 Miles to Find Out Why."

Brown's research involved stops on farms and visits to small town cafés, plus interviews with politicians, ag economists, and farm organization leaders. One of her more poignant experiences came in revisiting Downs, where in the 1980s she had attended the wedding of a close friend. The Smoky Hills town had a population of 1,324 back then, and its one-street downtown "teemed with life." Thirty years later, Downs had dwindled to 844 residents and most of its storefronts were empty. Brown wrote, with poetic overstatement, "That's the thing about rural Kansas: No one lives there, not anymore."

James E. Payne spent a good part of 1899 driving a wagon around the plains of eastern Colorado. Superintendent of the experiment station of the Agricultural College of Colorado (now Colorado State University) in Cheyenne Wells, Payne was tasked with making a reconnaissance of the

High Plains to study the agricultural methods of settlers "who had gained a foothold" in the challenging environment. He ended up traveling over thirteen hundred miles, most of it in the headwaters region of the Smoky Hill and Republican Rivers.

Payne reported his observations in *Investigation of the Great Plains: Field Notes from Trips in Eastern Colorado*. Under the heading "Depopulation," he named eight towns that had been established along the divide between the Arikaree River and North Fork of the Republican during the time frame of 1886 to 1889, when the area "was settled quite thickly." Of these, five had vanished entirely or had only a single family in residence. At one point he drove across eighteen miles of wide-open plains without seeing any human habitation, despite the fact that only a few years earlier nearly all the land had been homesteaded. The country had emptied out.

The woodlands and grasslands that grace the country of the Kaw are natural communities of plants and animals that have assembled and been refined over tens of thousands of years in response to the rigors of the Great Plains environment. The human communities of our region are much more recent experiments and, as Brown and Payne witnessed, are still working out their ecologies.

In his overview of cities and towns for the *Encyclopedia of the Great Plains*, University of Chicago geographer Michael Conzen identifies three archetypal urban expressions for the region. There is the gateway city that, like Kansas City, is the hub of transport and commerce for the hinterlands. Then there is the ethnic town like Wilber, Nebraska, where during the annual Czech Festival I witnessed an orchestra of eighteen accordion players performing snippets from a couple dozen Old Country polkas, in perfect unison and entirely from memory.

And, finally, there is the "plains country town"—the classic small town of the Great Plains. Conzen gives no parameters of population size for such a town, only that in its quintessential form it is "small enough to be walked through from end to end in a few minutes."

Merrill Gilfillan calls them kingbird towns. "A kingbird town is a phenomenon endemic to the Great Plains: small dry villages where on summer afternoons the only sound of life is the chatter and squabble of kingbirds in the wind-bent Chinese elms."

Gilfillan coined the term in his 1988 book *Magpie Rising: Sketches from the Great Plains*. Living at the time in Boulder, Colorado, the bird-loving poet and essayist traveled the plains from Montana to Texas, swinging into towns and villages along the way, notebook at the ready, to sample local life. He was especially taken with the smallest of these inhabited places, the ones that "occupy an ecological niche between the town and the isolate self-sufficient ranch-farm complex." The sound he kept hearing in these places was the fussing of the western kingbird, a handsome if noisy thing with a pale gray head and back and bright yellow belly.

It is not necessarily a bad thing to be a kingbird town. Many are just small hamlets that have always been small hamlets, and are quietly going about the business of life. Such places, typically with only a couple hundred residents, can have an understated dignity, even an elegant simplicity about them, like the elderly lady who is well aware of her age but wouldn't think of leaving the house without dressing up and doing her hair.

But many of these towns are in shambles, with Main Street storefronts not only empty but the buildings themselves caving in. Drive around a bit and you'll come across the old school, reluctantly closed and sometimes repurposed into commercial property or just boarded up. The town's churches will usually be well-kept, although you may find one that has been turned into someone's residence. Junked cars, trucks, and an assortment of other defunct mechanical equipment will decorate front yards and empty lots. Adding to the dystopian air, most of the trees in town will be dying Siberian elms. Often called "Chinese" elms, these were widely planted on the plains in the early 1900s because of their drought tolerance but are now at the butt-ugly end of their life spans.

It is sometimes hard to find a bite to eat in a kingbird town, even one that seems to be holding its own. If you're fortunate, you'll come across a café, tavern, bakery, donut shop, or, during summer, a drive-in dairy freeze. But sometimes the only food served is at a gas station convenience store. Whatever the venue, a step inside will earn you curious glances from the locals and provide a quick scan of rural community demographics.

I'm no sociologist, but I've walked through the doors of a lot of small-town eateries over the years. In the more distressed communities, the majority of patrons at the tables will be senior citizens, many of them in their seventies and eighties. And, sad to say, you'll likely spot a few folks with physical disabilities or struggling with obesity or who otherwise just wear the look of hard times, even some of the younger ones.

The populace of the most woebegone of these towns includes residents who just can't go anywhere else. The reasons might be economic (too poor to leave), vocational (lack of transportable job skills), emotional (too deeply rooted to leave), or a mix of these and others. But before we dig more deeply into these tragic circumstances, a bit of historical context is called for.

The places we're concerned about were founded by Euro-Americans in the middle to late 1800s or the very early 1900s. History, of course, runs far deeper than this, and the country of the Kaw was far from vacant before "settlement." There were horse nomads like the Cheyenne, Arapaho, and Lakota who, like the dog nomads before them, held sway over fiercely loved hunting territories. And semisedentary tribes like the Kansa, Pawnee, and Wichita, who had fixed villages and gardens on the eastern edge of the grasslands.

While these peoples faced the same arduous environment that challenges today's Great Plains communities, they had to deal with the ever-present danger of competing human cultures that coveted their homeland and were willing to kill to possess it. For centuries their adversaries were other indigenous peoples, but in historical times they were agents of the US government who by deception, coercion, or lethal force cleared the plains for stockmen, farmers, and town builders.

Whatever trials confront contemporary plainsmen and plainswomen, they pale in comparison to the trauma and sufferings of those who occupied the land before them.

As attested to by the travels of James Payne and Corie Brown, the overarching story of human communities in the Great Plains during the past hundred-plus years is a recurring cycle of boom and bust, good times and bad times, retreat and rebound.

This signature trait of the Great Plains experience is explored in *The Last Days of the Rainbelt* by David Wishart. The University of Nebraska geographer examined the history of Euro-American settlement in the central Great Plains, particularly in eastern Colorado, western Kansas, and southwestern Nebraska—the western reaches of the country of the Kaw.

While the devastations of the Dust Bowl years of the 1930s are stamped upon the American psyche, Wishart documents other rounds of settlement and collapse in the plains that occurred before and after this apocryphal event. The most dramatic was a boom during the second half of the 1880s that came to a crashing end by the mid-1890s, resulting in the vacant landscape James Payne saw from the seat of his wagon in 1899.

But that was not the end of the story. A number of communities in the region not only survived the 1890s but also started on a fresh trajectory of growth that peaked in the 1920s and early 1930s, when most reached their maximum population. Then the Dust Bowl happened.

The driving force behind these cycles has always been rainfall. Surges in settlement are coincident with years of decent or better-than-average precipitation that fuel optimism for the agricultural potential of the plains. Depopulation occurs when drought, a distinguishing and inescapable feature of Great Plains climate, brings crop failures.

The Rainbelt name was bestowed on the region by nineteenth-century boosters who adhered to or at least propagated the belief that "rain follows the plow," the idea that rainfall would increase on the western Great Plains as more of the shortgrass prairie was turned to farm ground. Trust in human mastery over the environment persists in the region yet today, primarily through the modern technology of irrigated agriculture. But, as Wishart observes, "Occupying the High Plains has always been a tenuous endeavor."

Whatever the backstory, the extraordinary smallness of the "plains country town" makes it seem almost alien to the vast majority of Americans who live in cities and suburbs. Taking the bus across western Kansas, the urbane narrator of Jack Kerouac's 1955 novel *On the Road* snarked about "crackerbox towns with a sea for the end of every street." As a teenage boy headed from Kansas City to Colorado on family vacation, I wondered about the girls who lived in such towns, as you might the girls who lived in a foreign country. These places might be no smaller than the classic New England village but, situated under the immense dome of a Great Plains skyscape, they seem little in every sense of the word.

The epitome of small-town smallness in our part of the world is the game of six-man football, which was invented and first played in the country of the Kaw. It originated in the town of Chester, Nebraska, where, during the depths of the Great Depression, there were no longer enough high school–age boys to field the standard eleven-man football team. But rather than give up on a beloved sport and vital symbol of hometown identity, local teacher Stephen Epler devised a set of modified rules that would allow for fewer players.

Using Epler's new rules, the first six-man game debuted on September 26, 1934, in the nearby town of Hebron. Born in the hard times of the 1930s, six-man football is played yet today in little plains towns from Canada to Texas. Thirty-two Nebraska schools fielded six-man teams in 2022, including those of the Rainwater Basin towns of Dorchester and Harvard, who I watched play each other one beautiful September evening in the company of my friend Wayne Howlett, a member of Dorchester's legendary undefeated six-man squad of 1960.

To many, small also means insignificant, irrelevant, expendable. I've had conversations with public sector and philanthropic leaders who view the survival of these towns as a matter best left to the play of Darwinian forces. But the most jarring point of view is the infamous proposal of Drs. Deborah and Frank Popper to essentially vacate the western Great Plains and return it to bison pasture.

In a paper titled "The Great Plains: Dust to Dust," published in the journal of the American Planning Association in 1987, the Poppers, then on the faculty of Rutgers University, described depopulation brought about by recurrent agricultural failures. They predicted imminent economic and environmental ruination that could only be prevented by intervention of the federal government. The aim of this massive effort would not be to help people stay on the land, but to buy them out and resettle them so that a Montana-to-Texas "Buffalo Commons" could be reestablished.

It took a while for the Buffalo Commons proposal to move out of academic circles and into broader public view, but once it did it sparked a firestorm among folks who actually lived on the plains. It didn't help that the Poppers called their way of life "the largest, longest-running agricultural and environmental miscalculation in American history." To have your world and family history critiqued and found wanting by academics from New Jersey only added insult to injury. But the most arrogant, and chilling, sentiment in the Poppers' paper was their pious call for "compassionate treatment for the Plains refugees" by the government agencies that would relocate them.

To their credit, the Poppers gamely took their vision on the road in 1990, making a fourteen-hundred-mile speaking and listening tour through the Great Plains. As described in *Where the Buffalo Roam* by Anne Matthews, the Poppers began the tour in the country of the Kaw. Their very first public meeting was in the southwest Nebraska town of McCook, where they spoke to a packed crowd in the high school gymnasium. The next day they headed toward Denver, swinging through Hayes Center and Benkelman on the way. In a field near Hayes Center the photographer hired by the *New York Times* to document the Poppers' travels had the couple pose, sardonically, atop a large round hay bale, the professors dressed in business attire and clutching attaché cases.

Anticipating trouble during the Poppers' appearance in McCook, the county sheriff had a half dozen deputies report for duty at the high school. Emotions did run high that evening, and one incensed audience member had to be escorted out of the building during the question and answer session. But the citizens of McCook, whose high school mascot is, ironically, the Bison, made some hay of their own out of the Poppers' visit. In 1997, they launched the Buffalo Commons Storytelling & Music Festival, still going strong today, and eventually branded their town as the Capital of the Buffalo Commons.

McCook has done pretty well for itself. It helps that the Republican River town of 7,555 residents sits at the junction of two major (albeit two-lane) highways, US 6/34 and US 83, making it a hub of commerce for a region that encompasses the corners of three states. Plus, it is the county seat and home to a community college, well-kept parks, and a diversity of businesses. But McCook is also a hotspot for entrepreneurs as well as arts and culture, exemplified by the legendary Sehnert's Bakery & Bieroc Café, which received the James Beard America's Classic Award in 2019, and a recently opened contemporary art gallery. McCook is the Queen City of the Republican valley and one of the most vibrant towns on the High Plains.

But not every town in the western Great Plains has the assets and energy of a McCook and, as the Poppers predicted, many are on the ropes. Yet even these are fighting to keep the lights on.

The surest sign of life in such a town will be found in the community park. No matter the state of disrepair in the business district, the grass in the park is often golf-course perfect—watered, fertilized, and groomed into an emerald carpet. And there is usually newer playground equipment—brightly colored tornado slides and other play structures instead of rusting teeter totters and swing sets—plus a few recently planted shade trees. I suspect the loving attention lavished on these places represents an act of defiance in the face of decay and, more heartbreakingly, a thinly veiled gesture to draw young families back home from Fort Collins or Overland Park or wherever else they've taken the grandkids.

You can't fault the residents of small towns for starting small. The forces faced by rural plains communities are monumental—a globalized economy, commodity prices shaped by the weather in Argentina, droughts triggered by ocean temperatures, aquifer depletion, access to quality health care, access to technology, to name a few. Thankfully, a host of state and federal agencies are working on these and other large-scale issues, as are the three land grant universities that serve the country of the Kaw—Colorado State, Kansas State, and the University of Nebraska.

But the most pressing need for the region is an agriculture that is both ecologically sustainable and economically profitable under the constraints and volatility of the Great Plains environment. On the ecological side, this means crops or cropping systems that use less water and require less input of fertilizer and pesticides. On the economic side it means agricultural products that earn higher profit or that can be marketed more directly to consumers, things like organically grown beef or wheat or other specialty niche crops.

While helpful, these innovations are mere tweaking compared to the revolutionary work under way at the Land Institute, a private nonprofit research and educational enterprise located on 650 acres of land along the Smoky Hill River near Salina, Kansas. The Land Institute's team of plant breeders, agronomists, and ecologists aim to develop an alternative to industrial agriculture that mimics natural ecological systems. The work involves breeding new perennial food-grain crops adapted to a form of cultivation called polyculture.

This Natural Systems Agriculture is the brainchild of Wes Jackson, who with his wife, Dana, cofounded the Land Institute in 1976. A Kaw valley farm boy with a PhD in genetics, Jackson looked to the prairie as the model for a more ecologically wise approach to agriculture. In contrast to a conventional monoculture built around the planting of a single species of a one-season (annual) crop like corn or soybeans year after year, a perennial polyculture is more like a pasture or domestic prairie that has been seeded to a mixture of relatively long-lived (perennial) crop plants. Less frequent cultivation leads to less soil erosion (and attendant siltation of waterways), and greater diversity of species in the mix offers the possibility of increased nutrient retention in the soil and reduced vulnerability to pests and disease.

Land Institute scientists are currently working to domesticate intermediate wheatgrass, a perennial relative of wheat, as a grain for human consumption. Marketed under the name Kernza®, the grain shows potential as a substitute for wheat in foods like baked goods and beer and can be used as a whole grain like barley or rice.

The folks at The Land estimate it will take another forty years to fully develop and implement Natural Systems Agriculture. As I write this in 2022, that means sometime in the 2060s. While this new paradigm holds tremendous promise as a sustainable form of agriculture for the Great Plains and other parts of the world, the residents of our struggling towns have nearer horizons in view. So, why not start by fixing up the park?

Money for park renovations and other community improvement projects is sometimes available from a government economic development grant or the special initiative of a large private foundation. But more often it comes from local contributions given to a local community foundation. The legal and fiduciary aspects of setting up and managing such a fund are complicated, especially when dealing with estate gifts and endowments. In Nebraska, the business side of things is facilitated by the Nebraska Community Foundation. Based in Lincoln, NCF holds and manages affiliate funds that serve 270 Nebraska communities, most of them small towns, about sixty of them in the country of the Kaw.

Jeff Yost is the president and CEO of the Nebraska Community Foundation and a great storyteller. I met with him to learn more about the work of NCF and to get his insight into the challenges that face rural communities. But he was determined to focus on success stories ("Good news is our super power"), of which there are multitudes. NCF affiliate funds have implemented an amazing variety of quality-of-life and economic development projects across the state.

Born and raised in Red Cloud, population 962, Jeff has worked with NCF since 1998 and knows a thing or two about the local narrative of small towns ("coffee-shop talk") and how it can sometimes drift toward deficits and pessimism. So, while NCF has plenty of financial people on staff, they also do a significant amount of coaching and training of hometown leaders, especially in "asset mapping." In Jeff's view, the flourishing of rural communities is not a matter of recapturing the past but of reimagining a new future and investing in projects that give the town a margin of excellence.

The wisdom of this approach was evident to *New York Times* columnist David Brooks after Jeff guided him on a quick tour of several rural Nebraska communities in 2019, including McCook. Brooks later wrote, "Everybody says rural America is collapsing. But I keep going to places with more moral coherence and social commitment than we have in booming urban areas."

The Kansas Sampler Foundation is another nonprofit that works on behalf of rural communities in our region. Headquartered on the family farm of executive director Marci Penner near Inman, Kansas, the organization orchestrates events, tours, contests, conferences, and other initiatives to encourage and facilitate exploration of the state. One of their most successful tools is the *Kansas Guidebook*. The latest edition, published in 2017, features forty-five hundred entries based on firsthand visits by Marci and her associate WenDee Rowe to each of the state's 627 incorporated towns and cities.

Marci is passionate about preserving rural culture, which she defines as the combination of eight elements: architecture, art, commerce, cuisine, customs, geography, history, and people. The Kansas Sampler Foundation helps bring to light unique restaurants, art galleries, historic sites, museums, scenic byways, festivals, performance venues, natural landmarks, architectural treasures, and other often below-the-radar assets woven into the fabric of rural communities. One of my favorite entries in the *Guidebook* is the "World's Largest Collection of the World's

Smallest Versions of the World's Largest Things" in a museum in Lucas, Kansas, a Smoky Hills town of 395 residents that attracts creative types who practice grassroots or "outsider" art.

Like their rural counterparts, many urban areas in the country of the Kaw have had their ups and downs over the years. The most distressed are neighborhood-scale places in the Kaw valley bottomlands of Kansas City, Kansas. The communities of Argentine, Armourdale, and Rosedale sprang up from their namesake bottoms during the boom times of the meatpacking industry, but devastating floods in 1903 and 1951 took a toll on their residential areas and business districts and eventually undercut the industrial economy that supported the mostly low-wage workers of racial and ethnic minorities who lived in the river wards. In Armourdale, for example, the population declined from a peak of twelve thousand in the early 1920s to fewer than twenty-five hundred today. These communities have since struggled with poverty, crime, substandard housing, and other woes.

But these neighborhoods also have their heroes—individuals and organizations that strive to make them healthy places to live and work. Daniel Serda is one of them. Born and raised in the Hispanic community of Armourdale, with degrees from Harvard and MIT, Dr. Serda has been a consultant and advisor to many community groups and revitalization efforts in Kansas City, advocating for diverse neighborhoods, thriving business districts, and the preservation and adaptive reuse of older buildings. He also has deep personal and historical knowledge of the bottomland neighborhoods and, through his publications and speaking engagements, has helped raise awareness of the key roles these places played in the rise of Kansas City as a major American metropolis.

After Kansas City, Kansas, the largest cities in the country of the Kaw are Topeka and Lawrence. Like KCK, both have historic river bottom districts that today are being revitalized. The neighborhoods of North Topeka and North Lawrence are located on the other side of the Kaw from the core of their respective cities, and both were devastated by the flood of 1951. As in the bottomland neighborhoods of Kansas City, some of the first pioneers in these urban renewal efforts have been creative types and young entrepreneurs who like the edgy, gritty ambience and can find an affordable place to set up shop in the well-worn building stock. This once-dispirited part of Topeka even has its own cool acronym—NOTO, the North Topeka Arts & Entertainment District.

The most audacious community revitalization scheme in the country of the Kaw involves the West Bottoms, an area that straddles the Kansas-Missouri state line near the confluence of the Kaw with the Missouri. In recent years, a number of scary old warehouses in the West Bottoms have been rehabbed into coffee houses, breweries, art galleries, antique markets, vintage clothing boutiques, even a whiskey distillery, making the area attractive to the young and the hip as well as suburbanites who enjoy the other-side-of-the-tracks vibe. But a bold new development is gaining steam that promises to make the Bottoms an even bigger draw.

The Rock Island Bridge spans the Kaw about two miles upstream from Kaw Point. Originally constructed in the 1880s to support the stockyards and meatpacking industry, and completely rebuilt after the 1903 flood, the bridge has stood idle since 1972 when the rail line it carried was abandoned. But a remarkable public, private, philanthropic, and corporate partnership has come together to repurpose the Rock Island Bridge into a one-of-a-kind entertainment district complete with restaurants, coffee shops, bars, event spaces, even gardens. The developers hope "America's first destination landmark bridge" will open in 2024. The whole thing feels wonderfully redemptive—an old bridge, built to bolster industries that contributed so much to the economic muscle of KC but that generated so much pollution, will now draw people back to the Bottoms for a craft cocktail and a view of sunset over the Kaw.

In his book *Cities on the Plains*, University of Kansas geographer James Shortridge identifies a list of assets that historically have given communities a competitive edge and better-than-average prospect of perseverance. But he also acknowledges that there are certain towns where, despite all odds, "a visitor gets the sense that a few key people simply willed their communities into continued life." Such folks are still at it throughout the country of the Kaw, from postindustrial urban neighborhoods to High Plains kingbird towns.

SIXTEEN

The Showalter Lilac

> Like that one great aunt in every family who takes it upon herself to remember the birthdays of every member, this old lilac has set about to chronicle the history of this farmstead.
>
> *Ted Kooser,* Local Wonders: Seasons in the Bohemian Alps

One of the ways Euro-American settlers sought to civilize their little bit of Great Plains wilderness was to plant "flowers." Of course, this happened after they broke the prairie sod and put in crops for cash and grain for their draft animals, and probably after they planted a vegetable garden and a few shade and fruit trees.

But the truest act of domestication on the frontier was the cultivation of plants that had no value other than just being pretty—garden flowers like peonies, iris, and daylilies and flowering shrubs like roses and lilacs. Yes, the prairie was teeming with beautiful wildflowers, but these were, well, *wild*. What the settlers desired was something familiar, something that carried a comforting sense of "back home." No plant was better suited to the task than the lilac.

I've poked around a lot of old farmsteads, schoolyards, and cemeteries throughout the Great Plains, and it is not unusual to find even the most run-down place still ornamented with lilac bushes. If these tough old shrubs could talk, and you could get them to open up to you, they would no doubt have tales to tell.

But of all the lilacs I've encountered over the years, one held history that I just had to pry into.

I came across the lilac in a cemetery to the south of Hesston, Kansas—a small, cedar-protected swatch of ground patched in among fields of wheat and milo. I had driven past this little graveyard a number of times but usually in a bit of mental fog, never expecting to see much of interest. Then one day I glimpsed the lilacs, the biggest I had ever seen, growing in symmetrical copses twenty feet in diameter. Scattered about the cemetery, each domed thicket was about five feet tall in the center, tapering to three feet at the perimeter. Hillocks of lilac.

I later learned this was the old Fairview Cemetery, established in 1872 next to the Fairview Meeting House, a place of worship for a local Quaker community. The meetinghouse and the tiny settlement that grew up around it eventually dissipated, leaving this lonely country cemetery as its only trace.

It was an afternoon in September, the first time I stopped. The spring lilac blooming season was long past, but the foliage was still dense and hearty in the late summer warmth. The grass had recently been mowed, and up against one headstone errant sprigs of big bluestem and sideoats grama revealed prairie underfoot.

My second visit was on a cold, clear day in the month of February. We would be moving from Hesston in a couple of weeks and I felt compelled to drive out and walk the cemetery, one last time.

The lilacs were dormant and the absence of leaves allowed me to study their architecture, like looking into the framework of a house being built. Each circular thicket owed its origin to a single shrub that had produced concentric rings of new shoots to its outside. Judging by dates in the cemetery, some of the lilacs had been sprouting new canes for decades.

Walking around the edge of the largest lilac, I could see dark shapes at its center. Looking closer I realized they were headstones, hidden on my first visit behind a veil of glossy foliage.

I fought my way into the middle and discovered part of a family enveloped within. Sarah Bausman, *Wife of David W. Bausman*, was there, born 1853, dying forty-four years later. With her were John, Charlie, and

Milton, three sons who failed to see their first birthdays. I found no other Bausmans in the cemetery, not even David.

I plumbed a second lilac and came upon an even more sorrowful story, one that haunted me for years.

There are twenty-some different species of lilacs, most of them native to Asia. The type in the Fairview Cemetery, and in almost every other cemetery in the Great Plains, goes by the scientific name of *Syringa vulgaris*, composed by Swedish botanist Carl Linnaeus in his 1753 publication, *Species Plantarum*. This particular species is sometimes called "common lilac," *common* being the English translation of Linnaeus' *vulgaris*. It is also called "Persian lilac," derived from what Linnaeus knew to be its natural range: "Habitat versus Persiam."

Linnaeus was partially right. Lilacs did indeed come to the gardens of Western Europe via Persia, having been cultivated in Constantinople, capital of the Ottoman Empire, as early as the 1500s. But in the late 1700s English travelers discovered natural stands of *Syringa vulgaris* on the Balkan Peninsula in what is today Bulgaria, and subsequent research has shown its natural range to be southeastern Europe, chiefly in the mountainous regions of Romania, Bulgaria, and Greece. So, at some point in horticultural history, lilacs were brought from the Balkans to the Sultan's Garden in Constantinople and from there were disseminated to the rest of the world.

Lilacs bloom in the spring, peaking in May but starting as early as mid-April. The individual flowers are small and tubular and packed by the dozens into four-to-eight-inch-long clusters called panicles. Colors range from the purples to pinks to white. Flowering lasts about two weeks, and after that the big lugs don't have a lot of horticultural value except to block the wind or screen out unwanted views.

A lilac in full bloom is a glorious sight, but it is the fragrance of its flowers that captures the heart and reverberates in the memory. Few plants stir nostalgia like *Syringa vulgaris*, and anyone who grew up around lilacs will be jolted back to childhood with a whiff from a shrub in bloom or a bouquet brought into the home.

Lilacs have long been esteemed by residents of the Great Plains for their ability to thrive in a cold, dry, windy environment. Such hardiness isn't surprising when you consider that the native habitat of this plant is the exposed, rocky slopes of mountains. Lilac toughness is especially evident in places like Fairview Cemetery, where the shrubs had been growing unattended for decades.

Many of the lilacs planted during the early days had been dug up from yards and farmsteads and carried by settlers out to their new home site. But lilacs were also sold by nurseries that began to pop up on the plains.

One of these was the Harrison Nursery Company of York, Nebraska, established around 1876 by Reverend Charles Simmons Harrison. The reverend was an enthusiastic grower and breeder of ornamental plants who seemed to sense a special calling to beautify the farmsteads of the plains. He was also an effusive writer, and in 1905 published *The Gold Mine in the Front Yard and How to Work It: Showing How Millions of Dollars Can be Added to the Prairie Farm*. In it he devoted an entire chapter to "The Lilac," calling the shrubs "among the hardiest things we have" and adding, "They are ready to do their utmost to cheer and enliven the home."

Reverend Harrison was one of the earliest lilac fanciers in the country of the Kaw ("I have imported quite a lot from France"), but he would not be the most fervent. That title belongs to Max Peterson.

Max Peterson and his wife, Darlene, live on a wheat farm in Perkins County, Nebraska. Their thousand-acre operation is located about fifteen miles north of the town of Grant on a stretch of High Plains tableland known as the Perkins Table. Tucked into the heart of the Peterson place are four windbreak-protected acres devoted to lilacs. When I first visited there in the 1990s, Meadowlark Hill Lilac Farm had one of the largest private lilac collections in the world, close to eight hundred cultivars on display, representing over twenty different species.

Each of the cultivars growing at Meadowlark Hill Lilac Farm has its own name, and each plant is labeled and cataloged accordingly. Shorthand for "cultivated variety," a cultivar is a specially selected plant with certain desirable characteristics that are perpetuated by horticultural practices that maintain those characteristics.

Let's say you raise a lilac from a seedling and once it grows large enough to produce flowers you realize it has a unique color or blooming time or some other nice attribute. You could grow plants from seeds produced by your unique variety, but the resultant progeny will have the genetics of two different parents and are unlikely to have the desired characteristics you hope to perpetuate. If you want to produce plants with the exact same traits, you have to propagate them from cuttings or buds taken from stems of the "mother" plant.

Although new lilac varieties sometimes arise spontaneously in a garden, they more often are the result of an intentional breeding program in which a person makes carefully controlled transfers of pollen from the flowers of one lilac variety to another. Either way, the discoverer or developer of the new cultivar has the privilege of giving it a name. They might go with something simple and descriptive, such as "Early Double White" or "Mauve Mist," but most cultivar names are more personal and have some emotion behind them.

Many entries in the official cultivar register of the International Lilac Society read like snatches of horticultural poetry, such as "Hanafubuki," Japanese for "blizzard of falling cherry blossoms," and "Aliya Moldagulova," which honors a World War II sniper heroine of the Soviet Union.

The register includes a "Max Peterson" lilac, named in honor of Max by fellow lilac enthusiast Ken Berdeen of Maine, as well as some bred by Max and named by him for family members and friends. It also includes a lilac named "Lourene Wishart."

Max was a good friend of Lourene Wishart, longtime resident of Lincoln, who passed away in 1989. Lourene was a big personality who did everything with gusto and flare, from raising toy Manchester terriers to supporting the Nebraska Cornhusker football team; her salty fanaticism earned her a profile in a 1984 article in *Sports Illustrated* magazine. But the woman's biggest passion in life was the lilac.

Lourene grew up on a farm outside of Lincoln that was homesteaded in 1850 by her maternal grandfather, William Roggenkamp. Mr. Roggenkamp had come to Nebraska from Germany, where he worked in the Ludwig Spaeth lilac nurseries of Berlin. It wasn't long before he was planting lilacs on his new place, which the family called Lilac Farm. Lourene and her attorney husband, Joe, continued the family devotion by planting an extensive collection of lilacs on their property in Lincoln's country club district.

Lourene not only cultivated lilacs, she also cultivated lilac enthusiasts. She promoted lilacs wherever she could and, as a founding member of the International Lilac Society, she attended the inaugural meeting of the organization in Rochester, New York, in 1971. Max told me Lourene arrived at the convention in a lilac-colored Cadillac (newly painted) driven by her chauffer, Otto, who was dressed in a matching lilac-colored suit.

My first visit to Meadowlark Hill Lilac Farm was in May 1998. My son Greg was with me and I have a photo of him, eleven at the time, standing next to Max in front of a towering shrub Max said was one of his very favorites, 'Krasavitsa Moskvy'—the "Beauty of Moscow."

The work of renowned Russian lilac breeder Leonid Kolesnikov, Max's treasure produces showy clusters of rose-pink buds that open up into brilliant white, double-petaled flowers. Kolesnikov released his new cultivar in 1947 to honor the eight hundredth anniversary of the founding of Moscow. This splendid lilac, now grown around the world, won Kolesnikov the Stalin Prize in 1952, an honor typically bestowed on physicists, composers, poets, and architects.

There are many touching stories of how lilacs lay hold of a person's heart, but none is more moving than that of Leonid Kolesnikov. As a young man of twenty-five, Leonid found himself on the front lines during the end of World War I and the early days of the Russian Civil War. In the spring of 1918, as he traveled about the war zone, he encountered many old, ravaged Russian estates. Amid the death and desolation, he found lilacs in bloom. The experience set the course for the rest of his life.

Back at Fairview Cemetery, as I peered into the second lilac, I could make out the silhouette of a small, double-headed marker, shaped like the stone tablets Moses is carrying down from Mount Sinai in the old Sunday school pictures. I pressed on in and read the names—*Judda* inscribed on the left, *Firman* on the right, *Children of J. W. and M. E. Showalter*.

Judda was just twenty-five days old when he died; Firman a year and nine months. Both boys were lost in October 1876, Judda on the seventh, Firman sixteen days later.

On my knees, engulfed in lilac tangle, I tried to comprehend the Showalters' tragedy. An ache was gathering in my chest when I saw the other headstone, smaller and more weathered. I had to press deeper, on my elbows now, to read its inscription: *Carrie*, four years, six months, twelve days. She died October first.

Carrie and Judda, lost to their parents in the span of a week; how they must have clung to Firman.

I hunted for *J. W.* and *M. E.*, within the lilac and without, but could not find them. The children's graves seemed lonelier, colder, after I called off the search.

We moved from essKansas to Nebraska in 1994, but I never could put the Showalter lilac out of my mind. Who were J.W. and M.E. and what caused the deaths of their children? Even more troubling, how could the parents not be at rest here, too, with the graves of their little ones?

After twenty years of wondering, I finally contacted the Harvey County Historical Museum in Newton to see if they could help. An able and determined archival volunteer, Ron Dietzel, took up the task and, after several months of combing through old newspaper articles, census documents, marriage records, and genealogy databases, Ron pulled the pieces together.

The children's parents were John W. and Mary Elizabeth (Schoonover) Showalter. They were married in Illinois in 1868 and moved to Kansas in 1871.

The cause of death was an illness known at the time as "inflammation of the mucous membrane." That is how it was described in a notice Ron discovered in the November 2, 1876, issue of the *Newton Kansan* announcing the passing of Firman. It reads in part: "Mr. and Mrs. John Schowalter [*sic*] have again sustained a grievous loss. For the third time in three weeks they have lost a child." I must admit, I teared up at seeing Firman's name in print after all these years. The Showalter's oldest child, Anna, then seven, contracted the disease but survived.

While we can't know with certainty, it is possible that the illness that took the Showalter children is what would later be called diphtheria. If so, this sad story is even sadder. Diphtheria is a highly contagious bacterial disease that causes inflammation that hinders breathing and swallowing and, if untreated, can be fatal. Young children are particularly susceptible, and the disease caused innumerable deaths among pioneer families. Thankfully, an antitoxin for this dread illness, known as "the strangler," was developed in the 1890s. Diphtheria or not, the disease that claimed Carrie, Judda, and Firman moved swiftly and ruthlessly.

Ron's diligent research also turned up a possible explanation for the absence of John and Mary from the Fairview Cemetery. Following the notice of Firman's death in 1876, the next documented date in the couple's saga is 1892, when John married an Emma Carter. We hear no more of Mary until 1896 when she filed for divorce from John "on the ground of abandonment and gross neglect of duty." At some point between 1876 and 1892, the marriage of J. W. and M. E. Showalter fell apart.

It is said that nothing is harder on a marriage than the death of a child. To lose three little ones would be terrible trauma. The agony of losing them, one by one, over a span of twenty-three days is beyond imagining. And, if the children did die of diphtheria, their end would have been awful. I wonder if the crushing grief ultimately snuffed out the bond of affection between John and Mary.

John and Mary lived out their final years in McPherson County, just to the north; John passed in 1902, Mary in 1913. While they may have planted the lilac that eventually enveloped their children's graves, neither one chose to be buried with them.

So, we now know something more of the Showalter story. My archivist friend Ron finds it heartwarming that Carrie, Judda, and Firman have in a way been reunited with their mother and father. But I'm still a bit unsettled.

We wouldn't know any of this if not for that lilac, and I just can't shake the feeling that I was called back to the cemetery, back during winter, when I could look into its twiggy heart, see the children's stones, speak their names, and feel the weight of their loss. It was the children's lilac that brought their story to light.

I had long thought about going back some spring when the lilac was in bloom, to see the color of its flowers and take in their fragrance. But Ron informs me the cemetery has since been renovated and the old lilacs removed.

I guess it doesn't really matter. The work of the Showalter lilac is done.

SEVENTEEN

Lithophilia

> I sometimes think that in the west we have produced only two serious works of art: the Clovis point... and that wonderful arch along the waterfront in St. Louis.
> *James A. Michener,* Centennial

The late Harvard biologist E. O. Wilson proposed the idea of *biophilia* in a 1979 newspaper essay, then expanded on his concept in a 1984 book by that title. Literally *the love of life*, Wilson described biophilia as the innate tendency of humans to affiliate with life, "to move toward it like moths to a porch light."

I hereby coin the term *lithophilia*, and declare the human heart is likewise susceptible to *the love of stone*. I offer as evidence the intimate intertwining of geology and art discoverable in the country of the Kaw.

I bought the book because I thought it could fill in some gaps in my mental map of the Great Plains. That it did, but it also led me down a rabbit hole I can't seem to climb back out of.

Central Plains Prehistory was authored in 1986 by Waldo Wedel, a man of Kansas Mennonite roots who became the preeminent scholar of Great Plains archaeology and ethnohistory. The Smithsonian Institution scientist was intrigued by the interaction of human cultures with the plains environment and chose the Republican River watershed as the ecological setting for his survey. I wasn't really interested in archaeology, but I did appreciate Wedel's detailed descriptions of climate, topographic features, plant and animal life, and other aspects of the region's environment.

It was through the book that I first learned of Jones-Miller, a carefully studied bison kill site located in the valley of the Arikaree River southeast of Wray, Colorado. Archaeologists working there in the 1970s, led by Dennis Stanford, also of the Smithsonian, unearthed the remains of nearly three hundred bison in an ancient ravine, butchered by prehistoric hunters after several mass kills. These were not our contemporary *Bison bison*, but the extinct *Bison antiquus*, a bigger, bulkier creature with much longer, scarier horns.

The extraordinary concentration of bison bones at Jones-Miller makes it one of the most important archaeological sites in the Great Plains. But just as significant is the wealth of worked stone artifacts that were associated with the bones. Stanford reported finding over 130 such chipped-stone objects, including scores of projectile points.

What astonished me was that these were manufactured from a variety of identifiable types of stone obtained by pedestrian peoples from known sources scattered across an expanse of land six hundred miles long and two hundred miles wide.

And, so began my tumble into lithics—lithic technology, lithic procurement, lithic transport, lithic caching, lithic workshops, lithic scatters. The more I learned the more fascinated I became.

Hunter-gatherers like those responsible for the bone beds at Jones-Miller were almost always on the move, leaving little to no trace of habitation that we recognize today. The primary evidence of their existence are the stone tools they used to dispatch, skin, and butcher their prey. These tools, especially the spear or projectile points, are for the most part distinctive in shape, size, and stylistic features that reflect unique design and manufacturing technologies.

Like botanists with their plants and zoologists their animals, archaeologists have developed a taxonomy or typology of chipped-stone tools that they use to classify artifacts and assign them to a particular cultural group and time period. The earliest of these is the Paleoindian period, which spans a time frame of 9,000 to 13,250 years ago. The major Paleoindian complexes represented in the central Great Plains are Clovis, Folsom, and Hell Gap. These names, derived from the locale where the first point of its kind was discovered, represent the earliest peoples of the country of the Kaw.

The projectile points that brought down *Bison antiquus* at the Jones-Miller site were of the Hell Gap cultural complex, dated at 9,500 to 10,500 calendar years before present. The Folsom culture is older, 11,700 to 12,800 years ago, but still associated with the hunting of this animal. Clovis is the oldest culture of all, dated at 12,700 to 13,250 years ago, and is associated with the hunting of an assemblage of large mammals referred to as megafauna, of which the mammoth is emblematic.

Paleoindian projectile points are essentially spear or lance points, weapons bound to a shaft for thrusting into the flesh of the prey animal. They are generally the shape of a narrow leaf in outline, widest below the middle and longer than wide, tapering to a sharp tip. The length of Clovis points ranges from two up to six inches. Folsom points are more uniform in length, generally around two inches, and typically have a more deeply concave base with prominent basal "ears." Hell Gap points average two-and-a-half inches in length and are distinctively shouldered, with a broad tip and a stem that tapers toward the base, which is straight or slightly concave.

The creation of these tools involved a process known as flintknapping. We don't have space to consider flintknapping in detail, but it amounts to systematically chipping and flaking away pieces from a chunk of stone until the desired size and shape are achieved. The early stages of flintknapping involve forceful, well-placed strikes from a hammerstone, but final stages require a much more delicate touch, like cutting facets into a gemstone.

Both Clovis and Folsom points are "fluted," possessing a central groovelike scar intentionally created by the stoneworker by removing a flake from the base of the point. In Folsom technology, the flute extends the entire length of the point, while in Clovis the flute extends no more than half the length. This groove may have facilitated binding of the blade to the spear shaft, or may have enhanced penetration of the blade through hide and muscle. Whatever its purpose, the act of fluting required incredible skill and only appears in the archaeological record during Clovis and Folsom times.

Successful manufacture of stone tools depends on high quality raw material that can be knapped or flaked to produce and hold very thin, sharp edges. The material is often referred to as flint, but archaeologists use the more geologically precise terms of chert or chalcedony. These materials are formed when highly mineralized water seeps into bedrock and certain minerals in the bedrock are replaced by microscopic crystals of quartz called silica. The resulting cryptocrystalline stone is said to be siliceous or silicified. It is dense, very hard, often smooth and waxy to the touch, and it fractures in a circular or shell-like pattern.

Silicified stone occurs in discrete nodules, lenses, or zones within the larger geological formation, and it must be extracted from the surrounding rock. In Paleoindian times this required quarrying by hand, using bones, antlers, and pieces of wood as digging tools.

The projectile points found mingled with bison bones at Jones-Miller were fashioned from tool stone (another wonderful term!) obtained from five identifiable lithic procurement sites. The main sites were Spanish Diggings in Wyoming, Alibates in Texas, and Flattop in Colorado. The thought of these prehistoric quarries captivated me, and I felt an urging to see them for myself. The quest took a number of years, but I'm glad I followed the calling.

Located in eastern Wyoming, Spanish Diggings is one of the largest and most impressive prehistoric quarry complexes in the world. The colorful name was coined by cowboys who came upon the massive rubble piles in the 1880s and figured the quarrying must have been the work of gold-prospecting Spaniards from the Southwest. The quarry sites are scattered throughout the Hartville Uplift, a range of hills located about twenty miles east of the Rocky Mountain front. University of Wyoming archaeologist Charles Reher wrote, "It is almost impossible to convey the sheer magnitude of these quarry complexes." Two types of tool stone were quarried here, Hartville chert and Cloverly quartzite, the latter grayish red to purple, the former in hues of yellow, orange, red, and brown.

Five hundred miles south of Spanish Diggings in the Texas Panhandle, the Alibates Flint Quarries National Monument is set in the rugged canyons and breaks of the Canadian River. Over seven hundred prehistoric quarry pits have been documented at the monument, with many more known to occur on private lands in the vicinity. The object of all this digging was the Alibates agatized chert, found in ledges of dolomite rock that cap the bright red shales and siltstones of the Quartermaster Formation.

Alibates tool stone is distinctive in appearance and is one of the most highly recognized lithic resources in North America. Projectile points made from it are colorful and often mottled, streaked, or banded with shades of red, blue, purple, brown, yellow, white, and black.

In the Pawnee Buttes region of northeastern Colorado, the High Plains tableland falls away in cliffs down to the Colorado Piedmont, a drop of about three hundred feet. Several buttes stand to the south of this dramatic escarpment, isolated remnants of the High Plains rising above the erosion-cut piedmont. Flattop Butte is one of these lonely outliers. Studies in the 1970s found about two hundred quarry pits on top of the butte.

The tool stone quarried from Flattop is chalcedony, which, because it occurs in rocks of the White River Group geological formation that caps the butte, is classified as a White River Group silicate. A typical piece of Flattop chalcedony is lavender gray with a dull luster and may be opaque or translucent. It is a shade of purple you sometimes see in a very angry thunderhead.

The other major tool stone represented at Jones-Miller comes exclusively from the country of the Kaw. Technically Smoky Hill silicified chalk, it is often referred to as Smoky Hill jasper. This stone occurs in outcroppings of the Smoky Hill Chalk member of the Niobrara Formation in north-central Kansas and south-central Nebraska, usually in ledges along waterways. The quarry sites, smaller in size and more scattered across the landscape, are not nearly as spectacular as the other three. Artifacts made from Smoky Hill silicified chalk are usually some hue of brown or yellow, sometimes described as caramel.

I recently had the opportunity to examine the chipped-stone artifacts that emerged from the Jones-Miller bone beds. Originally held by the Smithsonian Institution, they were later transferred to the Denver Museum of Nature & Science. Having wandered about the workshops of Spanish Diggings, Alibates, and Flattop, I was thrilled to see and actually hold the creations wrought from each of these mystical places.

Of the tool stone sources represented at Jones-Miller, Alibates occurs at the greatest distance from the site, close to three hundred miles as the crow flies, followed by Spanish Diggings at roughly two hundred, Smoky

Hill silicified chalk at one hundred, and Flattop at eighty. The coalescence of lithics from these far-flung places in a shallow draw of the Arikaree River raises all kinds of questions about the mobility patterns of ancient peoples. What were they up to?

One scenario is that the kill site was the work of a single band of hunters that traveled an annual circuit through the Great Plains, visiting various quarries along the way to gear up on stone. Another is that several otherwise-dispersed bands came together here and brought with them lithics sourced from their core territories. Or perhaps the stone traveled through an ancient network of lithic trading.

Archaeologist Steven Holen favors the first scenario. Former curator of archaeology at the Denver Museum of Nature & Science, Holen has studied Clovis mobility patterns in the central Great Plains based on the distribution of chipped-stone artifacts manufactured from known sources.

Tools fashioned from Flattop chalcedony are particularly well represented in Holen's study region, which leads him to suspect that Clovis people knew Flattop Butte well and made it a regular stop on migratory routes that extended up to 350 miles. Holen imagines that Clovis bands gathered at the butte to make tools in late summer or early fall, which would allow different bands to exchange goods, information, even mates.

While we can only speculate about the travel plans of Paleoindian peoples, archaeological evidence in the country of the Kaw reveals they covered a lot of ground. Artifacts fashioned from Spanish Diggings and Alibates stone show up in many places in the Kaw watershed; both materials are represented in artifacts discovered at the Kanorado Archaeological District on a tributary of the Republican River in far western Kansas.

But Flattop chalcedony seems to have traveled the farthest, particularly in Clovis and Folsom times. Folsom-age artifacts found at the Nolan Site in southwestern Nebraska, ninety-some miles east of Flattop, are mostly of this stone. Flattop Clovis points have been discovered as far to the east as the Diskau site in the northern Flint Hills of Kansas, 350 miles from Flattop Butte. Such distances are amazing when you consider this stone was transported on foot.

One of the most intriguing pieces in the puzzle of tool stone travel is the lithic cache, a cluster of lithic material that was intentionally set aside at some distance from the source area with the expectation of recovery and utilization at a later date. Caches sometimes contain finished projectile points or early-stage projectile preforms, but they more often consist of broad, flat blades called bifaces from which a number of tools could be manufactured. The bifaces themselves were reduced from the parent stone by knapping to make them easier to transport. Oval or lens shaped, bifaces are relatively thin but may be quite large, sometimes more than twelve inches long.

Lithic caches are such rare and important archaeological finds that they are given names. A number of famous ones have been discovered in the country of the Kaw, particularly in the western reaches, including the Baller, Busse, CW, Harman, and Walsh caches. The amount of lithic material found in these caches ranges from fourteen items in CW to over ninety in Busse.

A cache generally contains only one kind of lithic material. The Baller and Busse caches were found in the Republican River watershed and consist entirely of pieces of Smoky Hill silicified chalk. The CW cache was found near an upland playa in Lincoln County, Colorado, and all of its components are of Flattop chalcedony.

The Baller, Busse, and CW caches are recognized by archaeologists as Clovis age. The practice of lithic caching is most strongly associated with the Clovis culture, a strategy that makes sense given that these earliest human explorers of North America were just learning the landscape and had no idea of what lay ahead of them, particularly as they moved eastward onto the plains.

Tom Westfall, an enthusiastic and respected avocational archaeologist from Sterling, Colorado, has had a hand in discovering and cataloging a number of lithic caches from the Great Plains. In his view, most of these were probably "rainy day" caches—a supply of stone the hunters could return to if they didn't find suitable lithics in the unfamiliar territory into which they were pursuing game.

While their ultimate purpose was to put food on the table, you can't view a display of Paleoindian projectile points in a museum or in the home of a private collector and not see them as works of art. Douglas Bamforth, professor of anthropology at the University of Colorado, believes such objects were made by "highly skilled individuals who cared about their craft."

Bamforth reached this conclusion after years of studying chipped-stone artifacts from archaeological sites across the Great Plains, including the country of the Kaw. In his experience, these objects are almost always "more beautiful and technically sophisticated than was necessary for them to serve their utilitarian purpose."

In James Michener's sprawling novel *Centennial*, he imagines the thoughts of "the knapper," a twenty-seven-year-old man of the Clovis culture, as he fashions projectile points at a camp in the cliffs of Rattlesnake Buttes, Michener's fictional name for the Pawnee Buttes region of Colorado. After describing the meticulous process of striking a point, Michener pictures the flintknapper, who "enjoyed his work and knew it to be good," breaking into a wide grin of satisfaction as he hands the piece to his clan's most renowned hunter: "Men of a later day would have lathes at their disposal and electric drills and computers to assist them in determining slope, but they would produce nothing which in beauty, utility and perfect workmanship would match this Clovis point."

Michener the novelist saw artistic genius, even "love," in the Clovis point; Bamforth the scientist even more daringly proclaimed such objects to be "gratuitously elegant." Something in the minds and hearts of the stoneworkers led them to look beyond expediency and fashion works of art that satisfied the universal human impulse to create beauty—beauty that would be recognized by you and me, twelve thousand years later.

I've had the privilege of viewing an exquisite Clovis point in the personal collection of Nebraska avocational archaeologist Dick Eckles. Dick found it on a terrace along White Rock Creek, a tributary of the Republican River, in the Smoky Hills of Kansas. The point, just over five inches long, is perfect except for a small piece missing from the very tip. It was fashioned from purplish-gray chalcedony sourced from Flattop Butte, nearly three hundred miles to the west.

Crafted with as much virtuosity as a hand-cut gemstone, it was found with an assemblage of other more utilitarian Clovis lithics characteristic of a kill or processing event. The ancient environment of the White Rock Creek valley was apparently a big draw for big mammals; the bones of extinct species of bison, camel, horse, llama, and even sloth are found in the area, along with a cluster of seven mammoths.

There is no direct evidence linking the Eckles artifacts with the demise of these particular animals, but an area thick with megafauna would certainly have attracted Clovis people, equipped with their most beautiful and deadly stonework.

Of all the archaeological lingo I've picked up since Waldo Wedel lured me into his world, my favorite is "lithic scatter." The phrase refers to the debitage of small chips and flakes found at a place where a flintknapper, in the final stages of shaping a stone tool, put on its finishing touches.

In my line of work, I spend a fair amount of time looking down at the ground, and over the years I've stumbled across a handful of lithic scatters. Such encounters always give me pause. While the scale of human enterprise evident at a prehistoric quarry is staggering, and the artistry of a Paleoindian point takes my breath away, sifting through the tiny bits of stone in a lithic scatter stirs a quieter sentiment, a warming of the heart that feels strangely like friendship. The smaller the chip, the closer I am to the mind of the maker.

I first ran into Pete Felten in Oberlin, Kansas. Not the man, actually, but his work. I was staying in Oberlin while I did some botanical fieldwork in western Kansas and came across his *Pioneer Family* sculpture (composed in 1971) near the downtown area. The next day my travels took me west to Atwood, where I saw the *Atwood Buffalo* (1980) in front of the Rawlins County courthouse. Then, heading back east, I stumbled across *Young Girl with Book* (1984) in front of the Graham County Public Library in Hill City.

I sensed something providential was afoot when I did a quick internet search and discovered that Mr. Felten was from Hays, which happened to be my next overnight stop.

Peter F. Felten Jr. was born in Hays in 1933 and lives there yet today. Perhaps the most prolific sculptor to ever come out of the Great Plains, he has been carving since 1957 and works almost exclusively in native Kansas limestone. No one has done more to create art from the stony heart of the country of the Kaw than Pete Felten.

Pete's most well-known work is his monumental *Four Famous Kansans*, displayed in the rotunda of the Kansas State Capitol since 1981. The massive figures, each about eight feet tall and two thousand pounds, depict Amelia Earhart, pioneering woman aviator, Dwight David Eisenhower, commander of Allied forces during World War II and thirty-fourth president of the United States, William Allen White, Pulitzer Prize–winning journalist, and Arthur Capper, newspaperman and the state's first native-born governor.

The city of Hays and other nearby towns are replete with works by Pete, so many in fact that the local convention and visitors bureau has produced a "Finding Felten" guide with descriptions and photos of his works plus maps for locating them. A number can be found on the campus of Fort Hays State University, as well as the grounds of museums, historical sites, and an assortment of tourist attractions.

Pete also keeps a ramshackle studio and gallery near downtown Hays, where I was fortunate to find him one October afternoon, chipping away on a small bison sculpture.

Several area churches have commissioned work by Pete. There is the *Capuchin-Franciscan Friar* (1978) in front of the spectacular Basilica of Saint Fidelis in Victoria and *Saint Frances of Assisi* (1978) near the door of the modest Saint Francis Catholic Church in the tiny village of Munjor.

At Trinity Lutheran Church in Hays there is *Christ with Lamb* (2001). Echoing the scripture, "Come unto me, all ye that labour and are heavy laden," Jesus is standing with his arms at his side and his hands upturned and open in a gesture of welcome. Pete made the hands a bit oversized, a nuance meant to convey strength, and included a lamb leaning hard against the leg of the one who is mighty to save yet "gentle and lowly of heart."

Across the street from Saint Fidelis is Pete's famous *Pioneer Family* (1978), which honors settlers who immigrated to Ellis County, Kansas, from the Volga River region of Russia, the first families arriving in the Victoria area in 1876. Descendants of Germans who left their homeland in the 1700s at the invitation of Empress Catherine the Great to farm in peace on the steppes of southern Russia and Ukraine, these Volga Germans or "Germans from Russia" immigrated to America when political upheavals threatened their way of life in Europe.

On the left side of the composition are the heavily mustached father and two sons, on the right, the mother and two daughters; the two halves of the family face each other. They are dressed for winter cold—the mother with a shawl draped over head and shoulders, the father with a sheepskin cap, all with heavy overcoats. Their body language shows determination, but I detect a bit of bewilderment on some faces. And, while the overall tone is heroic, Pete added a heartwarming touch with the youngest daughter clutching a floppy doll by its arm.

So beloved is this work that Pete was commissioned by the American Historical Society of Germans from Russia to carve an exact replica for the grounds of their museum in my hometown of Lincoln.

The medium for much of Pete's sculpture, including *Four Famous Kansans* and *Pioneer Family* in Victoria, has been Silverdale limestone, a variant of the Fort Riley Limestone formation found in the southern Flint Hills that yields particularly dense, strong, fine-grained stone well-suited to carving. While Fort Hays Limestone of the Niobrara Formation outcrops extensively around Pete's hometown of Hays, Silverdale holds up much better under the elements.

Pete has also worked in Fencepost Limestone, more commonly known as Post Rock, which outcrops in the Smoky Hills. Fencepost Limestone occurs in rather narrow beds in the landscape, eight to twelve inches thick, so it doesn't lend itself to large sculptures. But it does have interesting variations in texture and color. Pete's most technically sophisticated creations are bas-relief pieces he has carved into slabs of Fencepost Limestone; working with the natural streaking in the stone, he fashions a darker background for the raised carving, achieving a sort of painting in stone.

I've now seen quite a few of Pete's sculptures, and have some personal favorites. Close to the top of the list is his *Pioneer Family* in Oberlin. This sculpture earned special mention in the 2019 book *Pioneer Mother Monuments* for its unique take on an emotion-packed icon. Located in the median of the main drive into the downtown area, the sculpture depicts a frontier father, mother, daughter, and son standing in a circle, each facing outward.

Author Cynthia Culver Prescott points out that most pioneer family monuments feature a father and mother standing shoulder to shoulder with their children in front of them, everyone facing forward. Pete has the father and mother standing back-to-back and holding hands with their daughter and son; the innovative approach makes the work visible from all sides while it also conveys a fierce sense of devotion and determination.

But my favorite is *Young Girl with Book*, placed in front of the library in Hill City. The girl, barefoot and sitting cross-legged on the ground, is staring intently into her book with a snoozing cat pressed against her back and a glass of lemonade (I'm guessing) at her side. Along each of the four sides of the square base Pete has carved a variety of small objects that might swirl in a child's imagination fueled by reading: a sailboat, a stallion, a teepee, and a medieval castle. Total serenity, captured in a chunk of limestone.

The notion of lithophilia came to me after a trek to the Jones-Miller locality in May 1992. Hoping to get a feel for the landscape that was the journey's end of all the prehistoric stonework, I headed southeast of Wray, Colorado, assuming there would be a historical marker to indicate the location of this important archaeological site. I found no marker but did come across a gathering of pickup trucks at the base of a bluff on the south side of the Arikaree River.

It turned out to be a group of men putting the finishing touches on a fence around a hilltop memorial to Robert B. Jones Jr. With them was Janie Jones, Robert's wife, who graciously visited with me and explained how her husband had uncovered the bison bone bed in 1972. He had been leveling off some land on the north side of the valley to increase the reach of their center pivot irrigation system. Robert died in a machinery accident on the ranch in the fall of 1991.

The fencing around the tract consisted of a single run of black chain hung in graceful arcs between stone posts. The posts had the bright, unweathered look of recently hewn rock, and by the color and cut I could tell they were of Fencepost Limestone of the Greenhorn Formation.

I told Janie Jones that I admired the posts and was familiar with their use in the Smoky Hills. She seemed pleased and a bit surprised that I recognized the stone and told me that the posts had been quarried near Jetmore, Kansas. I passed back through the area a few years ago and found that in 2003 Janie joined Robert on this blufftop resting place with its wonderful view of the valley.

Paleoindian hunters brought their very best stone to the ancient Arikaree River valley to secure food to sustain their people; stone that had been transported hundreds of miles on foot from the most renowned quarries of the Great Plains. *Ten thousand years later*, Janie Jones brought stone posts from a Kansas quarry nearly two hundred miles away to honor the memory of her husband, stone of the same type used by Pete Felten in some of his most beautiful and inventive works of art.

Lithophilia!

EIGHTEEN

The Greatest Day in the History of Beeler

> ... such cross-cultural, interracial empathy was a
> secret gift of time and place.
> *Craig Miner,* West of Wichita: Settling the
> High Plains of Kansas, 1865–1890

I was headed west on Kansas Highway 96, not far from the town of Beeler, when I whizzed past a roadside historical marker with the intriguing banner, "Homestead of a Genius." I flipped around, pulled up to the sign, and was surprised to read that George Washington Carver had once lived in the area. I knew the basic story of this extraordinary man, who rose from slavery to become an internationally acclaimed scientist and humanitarian, but had had no idea he spent time in western Kansas.

I took a gravel road south into the rolling hills and found the site of Carver's homestead, marked by a modest stone monument. I was struck by the expansive view and the solitude and wondered what life was like for a black man out in this corner of the Smoky Hills in the late 1800s. As it turns out, it was pretty good.

George Washington Carver was born toward the end of the Civil War, in either 1864 or 1865, on a farm in southwest Missouri. His mother's name was Mary, and she belonged to Moses and Sarah Carver. His father's identity is not known, though there is speculation he was a slave from a neighboring farm who died in an accident before the child was born.

The fatherless baby would soon be motherless. Bandits raided the Carver farm, abducted Mary and the infant George, and carried them away into Arkansas to sell. A man hired by Moses Carver to recover them succeeded in finding the baby, who the raiders had apparently abandoned when he fell ill with whooping cough, but not Mary. George and his older brother, Jim, who had been hidden during the raid, never saw their mother again. They continued to live on the Carver farm and from them took their surname.

The site of the Moses and Sarah Carver farmstead is now preserved and managed by the National Park Service as the George Washington Carver National Monument. Located near the town of Diamond, Missouri, the ecological setting is Ozark oak woodland and savanna. While the natural vegetation has been considerably altered from Carver's day, if you walk the trail through the monument you'll encounter a landscape wooded with oaks of various kinds, along with sycamore, walnut, basswood, sassafras, pawpaw, and persimmon. The original Carver farm had tracts of tallgrass prairie as well, and a nearby 163-acre parcel owned by the Missouri Prairie Foundation was recently named the Carver Prairie in his honor.

It was here that young George fell deeply in love with the natural world. In a brief recollection of his early life, Carver wrote, "I literally lived in the woods. I wanted to know every strange stone, flower, insect, bird, or beast." His particular fascination with plants led him to create a secret wildflower garden: "Day after day I spent in the woods alone in order to collect my floral beautis [sic] and put them in my little garden." He hid the garden from view because "it was considered foolishness in that neighborhood to waste time on flowers."

Possessed from childhood with "an inordinate desire for knowledge," George started attending the Neosho Colored School, a school for black children in nearby Neosho, Missouri, in 1877. Not long after that he moved to Fort Scott in southeast Kansas to pursue additional education. On March 26, 1879, he witnessed the brutal lynching of a black man named Bill Howard by a white mob and saw the horrific abuse of his corpse. It was an experience that haunted him the rest of his life. He fled Fort Scott before dawn the next day and eventually reached Olathe, Kansas.

In Olathe, Carver was taken in by a black couple, Ben and Lucy Seymour. There he continued his education and completed the sixth grade.

In the summer of 1880, he followed the Seymours west to Minneapolis, Kansas, in the Solomon River valley of the Smoky Hills. Carver seemed to prosper in the little town; he continued his schooling, purchased property, and started his own laundry business. It was in Minneapolis that he added the initial "W." to his name, after discovering there was another George Carver in town who had been receiving his mail. At some point the "W." was widened into "Washington."

Carver succeeded in piecing together a remarkable amount of formal schooling by the time he had become a young man, but he desired still more. In 1885, he applied for admission to Highland College in the northeast Kansas town of Highland. Accepted on paper, Carver's admission was promptly rescinded when he arrived on campus and the principal realized he was black.

Devastated, Carver sought to escape Highland but first had to earn some money. He found work in town with a white family named Beeler. The Beelers had a son, Frank, who had recently moved to western Kansas and established a store that had become the nucleus of Beelerville, the town name later shortened to Beeler. There was still government land available for homesteading in the area so, in the spring of 1886, perhaps not really knowing or caring what he was getting himself into, Carver joined Frank Beeler out on the plains.

The 160-acre tract of Ness County, Kansas, that eventually became Carver's homestead is not prime farm ground. A contemporary of Carver would later comment, "I remember that quarter section well. The land was not so very good." The soil is rather thin, in some places downright rocky, and there are no creeks or springs on the property. Situated above the South Fork of Walnut Creek, the site is elevated enough that today you can clearly see the grain elevator in the town of Dighton, sixteen miles to the northwest.

Surely Carver, dubbed the "plant doctor" as a child for his mystical way with plants, could grasp the agricultural limitations. I wonder if, at this time in his life, he found solace in just having a bit of land he could call his own. He would eventually construct a fourteen-by-fourteen-foot sod house on the place.

It was early May when I walked Carver's homestead site for the first time, and the grassy hills were spangled with wildflowers. My field notes for the day list twenty-five different species in bloom in the mixed-grass prairie matrix, including bitterweed, plains penstemon, narrow-leaf purple coneflower, scarlet globe-mallow, plains skullcap, and Fremont's evening primrose. Carver officially took residence on his homestead on April 20, 1887, and lived there until June 1888, so he would have had two springs to wander his exuberant prairie garden. I'm sure he enjoyed the sight, but this child of the Ozarks was no doubt heartsick for Mayapple, bloodroot, and other joys of a woodland spring.

While Carver was captivated by plant life, he also had a keen interest in geology. "Rocks had an equal fascination for me," he wrote of his childhood days in Missouri, and he brought that curiosity with him to western Kansas. O. L. Lennen, editor of the *Ness County News*, recalled once finding Carver "wandering over the country near Beeler," adding, "on this day he was studying geology." Carver told his biographer Rackham Holt of two geological forays in the country of the Kaw, one to see Rock City near Minneapolis, where over two hundred giant sandstone orbs are scattered across a prairie pasture, the other to see Castle Rock, a spectacular chalk monolith in the valley of the Smoky Hill River.

Carver's intimate familiarity with the natural world as a child and as a young man made it easy for him in later years to resonate with the emerging science of ecology. The notion that living things are dependent on their surroundings, including other living things, and that Nature is the functioning whole of these interrelationships, was deeply known and felt by Carver. But in his mind, human beings were as integral to the whole as rocks and flowers, and behind all the wondrous complexity was an omnipotent yet personal God. Carver's love of Nature and Nature's Creator-Sustainer, coupled with the poverty of his early life, laid a foundation for the ethic of conservation and stewardship that would later pervade his teaching and writings.

Aided by his connection to Frank Beeler, Carver pieced together enough work in the area to make a meager living. Their friendship also provided him entrée into pioneer social circles. Able to play the piano, organ, accordion, guitar, and mouth harp, Carver, along with Beeler, frequently provided entertainment to settlers in the area. Further evidence of Carver's acceptance was his involvement in the Beeler Literary Society, of which he was elected assistant editor, a seemingly small honor but a distinction one historian called "astounding" given Carver's race.

Carver moved to the Beeler area a decade after the founding of the town of Nicodemus about a hundred miles to the north. Nicodemus was established in 1877 by Exodusters—African Americans from the South who trekked to Kansas for the opportunity to acquire public land. Situated along the South Fork of the Solomon River, the colony was envisioned as a place to start a new life, but drought, grasshopper plagues, and a railroad connection that never materialized conspired to take the shine off this "promised land." At about the same time, southern blacks were also settling near the confluence of the Kaw and the Missouri in Kansas City, clustering in enclaves of shanties with names like Mississippi Town, Juniper Bottoms, and Rattlebone Hollow. Carver was better off in Beeler.

Yet bigotry was not far below the surface. His first employer was George Steeley, whose property adjoined what would later become Carver's homestead. Carver was hired by Steeley's mother, who was visiting at the time, to be a general handyman around the operation. She recognized a good worker in Carver but was intolerant of his race and would not allow him to partake of meals with the family. The younger Steeley did not share his mother's prejudice, and, after she departed, he and Carver took their meals together as a matter of course.

While Carver's domestic skills and work ethic earned him employment, his keen mind and joyful inquisitiveness made him a curiosity and something of a local celebrity. The *Ness County News* published a profile of Carver on March 31, 1888, noting, "His knowledge of geology, botany and kindred sciences is remarkable and marks him as a man of more than ordinary ability." The article concludes, "He is a pleasant and intelligent man to talk with and were it not for his dusky skin—no fault of his own—he might occupy a different sphere, to which his ability would otherwise entitle him."

That the local newspaper would print a favorable story about a black man is quite surprising for its time, as was the sympathetic recognition of the injustices he faced. But these virtues are tempered by the almost casual resignation to the implications of racism.

Carver left Ness County sometime after June 1888. The next documented date in his life is September 9, 1890, when he enrolled at Simpson College in Indianola, Iowa, to study piano and art. The following year he transferred to the nearby Iowa Agricultural College, now Iowa State University, to pursue his passion for botany. Over the next four decades he would emerge from obscurity to become an internationally celebrated scientist at Alabama's Tuskegee Institute.

Folks in western Kansas apparently lost track of Carver, until Ness City banker George Borthwick came across a magazine article about him. Borthwick wrote Carver at Tuskegee asking if he was the man he knew from homesteading days. In a letter dated October 16, 1932, Carver replied, "How your letter astonishes me. Yes I am the same Geo. W. Carver. My, how I would love to see you and the other dear boys as well as some of my old haunts."

George Washington Carver died on January 5, 1943, and the nation mourned the passing of a great American. Carver was eulogized for his contributions to science, his service to struggling black farmers, and his role in tearing down racial stereotypes. More recently, he has been recognized as an environmental visionary and pioneering proponent of sustainable agriculture. But when word of Carver's passing reached residents of Ness County, Kansas, he was simply remembered as one of their most remarkable neighbors.

Sometime after Carver's passing, members of the Ness County Historical Society began to consider how best to memorialize his homesteading years. It took a decade, but on Sunday, October 11, 1953, a memorial marker was dedicated at the site of his former homestead.

The marker is a bronze plaque affixed to a boulder of greenish opaline sandstone. This native stone, which occurs locally in the Ogallala Formation, was a fine choice given Carver's love of rocks and penchant for collecting geological specimens in the area. Even more fitting, the monument site itself is fenced off from the adjacent pasture using the signature limestone fence posts of the Smoky Hills. The plaque reads: "Dedicated to the memory of George Washington Carver, 1864–1943, citizen, scientist, benefactor, who rose from slavery to fame and gave our country an everlasting heritage. Ness County is proud to honor him as a pioneer. This stone marks the northeast corner of the homestead on which he filed in 1886."

The dedication ceremony evokes wonderful images. A Mrs. Fayette Brown of Ellsworth heard about plans for the event and offered her services as a gospel singer, suggesting she could perform one or more "of our beloved Negro spirituals." She was given a place on the program as was the choir of the African Methodist Episcopal Church of Great Bend. Martha V. Robinson of Kansas City, a graduate of the Tuskegee Institute, gave remarks on behalf of the university and shared personal remembrances of Carver, including his "five a.m. daily talks with his flowers in his green house."

Mrs. Brown sang the national anthem and performed two solos. The choir sang two songs and, while the titles weren't noted, the *Ness County News* reported their performance added "variety and spice" to the program.

Special effort was made to invite African Americans from other western Kansas communities to the dedication. Editor Lennen of the *Ness County News* wrote an invitation to W. L. Sayers, a black attorney from Hill City, stating, "We shall be pleased to have you, and other members of your race especially, to be present on that occasion." The effort seems to have been successful, for the *News* reported that "colored visitors" from eight different Kansas counties were present. Total attendance at the dedication was estimated at fifteen hundred to two thousand people. In a recap of the event a few days later, the *Ness County News* declared, in a grand yet oddly guarded way, "This was probably the greatest day in the history of Beeler."

Carver spent only two of his almost eighty years on his patch of Smoky Hills prairie. But coming on the heels of rejection from college because of his race, those years may have been a crucial, healing episode in his life. In his book, *West of Wichita: Settling the High Plains of Kansas, 1865–1890*, historian Craig Miner observed, "Competition for population led also to at least temporary tolerance for minority groups in western Kansas that was remarkable in contrast to the general national tone. It was part of a push for unity, an aspect of hard surroundings where one took help from wherever it came." The kind of acceptance and even empathy experienced by Carver in the neighborhood of Beeler was, as Miner eloquently put it, "a secret gift of time and place."

Robert P. Fuller, past historian at the George Washington Carver National Monument, offered his thoughts on the importance of Carver's homesteading years in a 1960 interview in the *Salina Journal*: "It seems that it was a necessary period of quiet re-evaluation or thinking things through. It probably contributed to Carver's later achievements in much the same way that a period of lying fallow contributes to the productivity of a field."

It is tempting to end on these heartwarming notes. As a botanist, I love the thought of George Washington Carver—a man who admitted to talking with plants—striding across his flowery pastures, reveling in the beauty and in his freedom. But the story of Carver's prairie sojourn would be hollow without considering the malevolence that drove him out to western Kansas.

One of Carver's most cherished possessions was the bill of sale documenting the purchase of his mother, Mary, by Moses Carver. Mary was a thirteen-year-old girl on October 9, 1855, when she became the property of Moses Carver; the document warranted her "to be sound in body and mind and a slave for life." Historians suspect Mary was purchased from a neighbor, but there is also the possibility that she was bought at a slave auction or from a slave trader. Regardless, the terror of a young girl being sold to another human being is beyond imagining, as is the agony of the mother and father who were powerless to protect their child. Such was the brutal story into which Carver was born.

Slavery was abolished in the United States while Carver was still a babe, but the evil calculations that once made it thinkable shadowed him like a wraith through Kansas, from the lynching in Fort Scott to the rescinding of his admission to Highland, tainting even the benevolence of his homesteading years.

George Borthwick was a sincere admirer of Carver. The Ness City banker and civic leader gave the young black man a loan that enabled him to pursue his dream of college, and he seems to have done more than any other person to bring attention to Carver's time in western Kansas. Borthwick's high regard for Carver is evident in an interview printed in the *Kansas City Times* in 1942, in which he said that as he got to know "the young Negro" he "became more and more impressed with his extreme intelligence. He fairly seemed to glow with enthusiasm and his wide variety of information was amazing."

Borthwick went on to say, no doubt intending praise but revealing the still-entrenched bigotry of the day, "when I was in the presence of that young man Carver, as a white man, of the supposed dominant race, I was humiliated by my own inadequacy of knowledge, compared to his."

I came across the newspaper article in a box of Carver documents held by the Ness County Historical Museum, housed in a handsome Fencepost Limestone building in downtown Ness City. I visited the museum a couple of years after stumbling across Carver's homestead site and was graciously aided in my research by Rex Borthwick, great-grandson of George. I hesitate to quote the elder Borthwick given his kindnesses toward Carver, except that his words expose an impulse to prejudice that is hard for humanity to shake, a sin that has circulated in my own family over the years, and that I've had to tamp down in my own heart.

Little bluestem would have been in full autumn glory on the day the memorial marker was dedicated at the homestead site of George Washington Carver. Speeches were made, poetry was read, and the Boy Scouts raised the flag. But I like to imagine the singing: "beloved Negro spirituals" ringing out across the copper-red prairie! A properly sublime moment for the man who loved so deeply Creation and the Creator.

The ceremony no doubt caused many in the gathering to reflect on what makes a person truly great. If it caused some to consider what makes a person truly human, then it was indeed the greatest day in the history of Beeler.

NINETEEN

Rock Towns

> After the thrall of the grassland itself, the thing that lured
> me here was the stone architecture; the adroitly laid rocks
> of the courthouse, the Cedar Point mill, and the bridges,
> banks, homes, fences, cattle chutes.
> *William Least Heat-Moon,* PrairyErth: A Deep Map

William Least Heat-Moon came to the Flint Hills for a literary challenge: "I aimed to write about a most spare landscape, seemingly poor for a reporter to poke into, one appearing thin and minimal in history and texture, a stark region recent American life had mostly gone past." He set up shop in Chase County, Kansas, just south of the watershed of the Kaw, and began work on his epic *PrairyErth*.

Initially smitten by glorious prairie vistas, Least Heat-Moon found his head turned by another charm of the Flint Hills countenance—stone. He was especially taken with the courthouse in Cottonwood Falls, built in 1872 of locally quarried Cottonwood Limestone, laid down in Permian seas. He writes, "The building has been cut from the hill it stands on, and perhaps no other courthouse in America has traveled so short a distance from bedrock to hall of justice."

The Cottonwood is one of the most extensively used building stones in Kansas, the substance of such prominent structures as the state capitol in Topeka, the Memorial Campanile bell tower on the University of Kansas campus, and St. Marys' church atop Strawberry Hill in Kansas City, Kansas. But there is something about its service in the Chase

County Courthouse that is especially stirring. I think it is the profound particularity of it all—native stone, "the bones of the land" as Least Heat-Moon put it, quarried locally and used locally to express very local hopes and dreams.

The country of the Kaw has a wealth of native stone architecture, particularly in the Flint Hills and Smoky Hills. Take a road trip through these regions, swing into town squares and business districts as you go, and you will find that the really important buildings, where the business of life is transacted, are made of stone—the courthouse, town hall, post office, library, schools, churches, banks, park shelters, and livery-stables-turned-museums. And almost without exception the stone will be a locally quarried stone with a redolent local name like Cottonwood, Shellrock, or Fencepost. These towns are rock towns, anchored to place in the deepest sense of the word.

No matter where you start or finish, your rock town ramble will give you a taste of small-town charm and sometimes quirky vernacular architecture. Many of these buildings are not only beautiful but also cultural treasures, often on the National Register of Historic Places. *The Kansas Guidebook*, published by the Kansas Sampler Foundation, is an indispensable aid to finding these architectural gems.

But orchestrate your itinerary to take you from east to west, and you will add a chronological dimension to the excursion, experiencing on the horizontal what is stacked vertically below. I wonder if there is another region in America where the unfathomable vastness of geologic time is so nonchalantly benchmarked on Main Street.

The imprint of local geology on local architecture first showed up on the farm. Arriving in a land of little timber, and without the financial resources to purchase the lumber needed to construct houses and barns, the first Euro-American settlers embraced native stone as a building material. Such was Howard Ruede, who rode into the Smoky Hills settlement of Osborne City, Kansas, on a freight wagon in March 1877. The Pennsylvania-born Ruede wrote in his diary, "Stone can be had for the asking; or by going upon government land we can get it without asking. And a man can soon learn to be a stone mason here." He wrote a few weeks later, "I am going to have stone house in two years—not live in a dugout all the time, like a good many folks here." These early stone structures often reflected the unique architectural traditions and tastes of the settlers' homeland, particularly those with Czech, Scandinavian, English, or German roots.

Having secured habitation on the farm, the early settlers turned to town building. Many of the native stone structures in towns and villages in the country of the Kaw date back to the post-frontier boom times of the late 1800s and early 1900s when local populations and economic aspirations were on the rise.

Using county courthouses as a metric, this era is bracketed by 1884, when the Potawatomie County Courthouse in the Flint Hills town of Westmoreland was built, and the decade of 1900 to 1910, when many of the fine native stone courthouses in the Smoky Hills first opened their doors. Likewise, a host of stone churches were also constructed during this era, as congregations grew in size and prosperity.

But by the beginning of the 1920s, post-frontier regional development had mostly passed its climax in the central Great Plains, as had construction with native stone in the country of the Kaw. One factor was the availability and affordability of easier-to-use building materials like lumber and brick as transportation within the region improved. Another was a shrinking pool of laborers, as more men took work away from the farm. On top of it all, the old stonemasons were dying off.

A second wave of native stone construction hit in the 1930s, this one sparked not by an economic boom but by the double gut punch of a worldwide financial crisis and a national environmental disaster.

So, let's take that road trip, rolling east to west from the Osage Cuestas out to the High Plains. All of the towns and counties we'll be visiting are in Kansas, except for a handful in Nebraska that will be so noted.

In case you were only half awake during the first three chapters, let me remind you that the surface geology of the country of the Kaw is dominated by sedimentary rock formations: sandstones, shales, and limestones. Of these, limestone formations have seen the greatest use as building stone. As newspaperman Horace Greeley wrote in 1860, "The limestone itself is among the chief blessings of Kansas."

Within these rock formations, the lithology of certain layers or "members," that is, their color, texture, grain size, hardness, and durability, makes them suitable for use as structural stone and imparts a distinctive character to local architecture.

The Oread Limestone is the most important building stone cropping out between Kansas City and the Flint Hills. This rock caps the Oread Cuesta, a ridge that runs through the vicinity of Lawrence. Many buildings on the KU campus, which sits atop Mount Oread, are constructed from the Toronto Limestone member of the Oread formation, including two of the most venerable, Spooner Hall and Dyche Hall, the latter housing the Museum of Natural History. The Douglas County Courthouse in downtown Lawrence was also built of Oread Limestone.

Geologists recognize over twenty different limestone formations running the length of the Flint Hills, but three stand out for their importance as building stone—the Cottonwood, Funston, and Fort Riley.

The Cottonwood is the queen of the Flint Hills limestones. Named for the Cottonwood River valley in Chase County where it was first described, it has been used widely both within and without the Flint Hills. It was quarried so extensively around Manhattan that it was once called "Manhattan stone"; it was used in most of the buildings on the campus of Kansas State University, including the football stadium with its castle-like battlements. The Riley County Courthouse in Manhattan is also a Cottonwood building, as is the Clay County Courthouse in Clay Center.

Tucked away in the valley of Mill Creek in the heart of the Flint Hills is the picturesque village of Alma, known as the City of Native Stone, where many of the historic downtown buildings and several fine churches were constructed from locally quarried Cottonwood Limestone. The Cottonwood was also used extensively in the northern Flint Hills town of Blue Rapids, including the public library, built in 1875 and still in use today.

The Funston Limestone was named for Camp Funston, a training camp on Fort Riley Military Reservation where this limestone is exposed in bluffs along the Kaw River. But it outcrops more extensively to the northeast around the town of Onaga in Potawatomie County, where it has been quarried under the more widely known name of Onaga limestone. Many buildings in Onaga were built from this limestone, including the quaint Saint Vincent DePaul Catholic Church. It was also used in the Dwight D. Eisenhower Presidential Library and Museum in Abilene.

The Fort Riley Limestone is exposed extensively in the western part of the Flint Hills, especially in the area of Junction City and Fort Riley. Named for old Fort Riley, a frontier US Army outpost established in 1853 to protect travelers on the Santa Fe and Oregon Trails, the limestone forms strong bluffs and rimrock along the Kaw River on and near the military base. The old historic buildings of Fort Riley were all constructed from this limestone, including officer's quarters that once housed General George Armstrong Custer and his wife, Libbie. More modern uses include the Geary County Courthouse in Junction City.

Exposures of Fort Riley Limestone in the northern Flint Hills have been quarried under the commercial name of "Junction City stone." A variant that crops out in the southern reaches of the Flint Hills, beyond the country of the Kaw, yields a building stone that is particularly dense and fine-grained in texture. Quarried under the name of Silverdale limestone, this localized expression of Fort Riley Limestone has been widely used in Kansas and beyond, its superior lithology lending itself to fine stonework and sculpting. As an example, the exterior of Christ Episcopal Cathedral in downtown Salina, a gothic revival gem built in 1906, is faced in Cottonwood Limestone, but Silverdale was used for the dressed stone that accents the windows and entries.

Fort Riley Limestone crops out sparingly in the Big Blue River country of southeastern Nebraska. You won't find a lot of stone buildings in this region, but there is one that is celebrated for its splendor and its importance in local history.

Elijah and Emma Filley left Illinois in 1867 with their two sons and homesteaded in Gage County, Nebraska. Elijah became a leading farmer and livestock producer in the area and eventually one of its most prosperous and prominent citizens. In 1874, the combined blows of a national financial panic and local crop failures brought on by drought and grasshoppers had many area farmers in desperate straits and thinking of heading back East. In response, Filley decided to build a barn, a grand, three-story stone barn, and hired men from all over the region to do the work.

Construction of Filley's barn in 1874, which included quarrying and hauling stone from a nearby outcropping of Fort Riley Limestone, provided employment that helped many of these men weather the hard times and stay on their farms. It was like an early version of a Depression-era work relief project, except privately funded. The Filley Stone Barn stands about ten miles east of Beatrice and is maintained by the Gage County Historical Society. This beloved local landmark is on the National Register of Historic Places.

Heading west into the Smoky Hills, you encounter three distinctive subregions, each dominated by a successively higher and younger group of erosion-resistant sedimentary rocks—the Dakota Formation on the east, Greenhorn Limestone in the middle, and Niobrara Formation on the west. Each of these subregions has rock units that are distinctively expressed in local architecture.

The only sandstone that has seen much use as building material in the country of the Kaw comes from the Dakota Formation. In certain exposures of Dakota sandstone, the grains are cemented by iron oxide, which makes the rock relatively hard and resistant to erosion and gives it an attractive rusty-brown coloration. You can find nice Dakota sandstone buildings in Ellsworth, and in the historic districts of Brookville and Steele City in Nebraska.

The most distinctive stamp of native stone on the cultural landscape of the country of the Kaw is found in the heart of the Smoky Hills, where the Greenhorn Limestone formation dominates the surface geology. Two subunits or beds of the Greenhorn have been utilized for building stone in this region, the Shellrock Limestone and the Fencepost Limestone.

Shellrock Limestone, geologically older and lower in the rock column, crops out in the eastern portion of the Greenhorn zone. It has been used most extensively in and around Concordia. The Cloud County Museum in Concordia, formerly the library, is built of Shellrock. Not to be missed, Our Lady of Perpetual Help Catholic Church in Concordia creatively blends Shellrock and Fencepost Limestone in a pleasingly eccentric castle-like structure.

Fencepost Limestone, the youngest and uppermost subunit of the Greenhorn, occurs in the landscape as an eight-to-nine-inch bed of rock. Early Euro-American settlers used this rock layer to fashion four-sided columns to serve as posts to support wire fencing, hence its name. While Fencepost Limestone is the geologically correct name for this unit of the Greenhorn, it is more commonly known as post rock and its region of occurrence as Post Rock Country.

When first uncovered, this chalky limestone is soft and relatively easy to quarry and cut into slabs of uniform thickness. Once cured, the stone becomes much harder and eventually weathers to a light tan color. The Fencepost bed typically has a two-inch-wide brown streak of iron-rich minerals running through the center, which gives structures built of this stone a slight butterscotch hue. The initial softness of the quarried stone makes it easy to work and allows for artistic stonemasonry touches like pitch-faced chiseling.

Three towns in Kansas—Beloit, Lincoln, and Russell—are set in the heart of Post Rock Country and are rich in native stone architecture. Each is a county seat, graced with a marvelous Fencepost Limestone courthouse in the center of town as well as all manner of other civic and commercial buildings.

It is safe to say that you will find at least one church built of Fencepost Limestone in every town and village within its outcrop zone, some humble, some extravagant. But to experience the most amazing entwining of post rock and piety in the country of the Kaw, head to the vicinity of Hays, where seven Catholic churches are clustered in seven nearby towns, each built of Fencepost Limestone and each an architectural delight in its own right.

The largest of these is the Basilica of Saint Fidelis in Victoria. Widely known as the Cathedral of the Plains, this massive structure has seating for eleven hundred people, which at the time of its dedication in 1911 made it the largest church west of the Mississippi. Next is Holy Cross in Pfeifer, a dazzling gothic style cathedral with three spires, the central one towering 165 feet above the tiny village. Saint Joseph in Liebenthal and Saint Anthony in Schoenchen are smaller but likewise dominate their towns. The other three, more modest, post rock churches are found in the towns of Gorham, Munjor, and Walker.

These churches were built between 1890 and 1918 by congregations of Volga Germans who immigrated to this part of Kansas from the Volga River region of Russia. Arriving in the area in the late 1870s and 1880s, these devout Catholic farmers began building churches as soon as they could, initially small wood-framed structures but later, as they grew in numbers and prosperity, more permanent buildings of stone. With beds of Fencepost Limestone close at hand, abundant volunteer labor from parishioners, and a man or two in the community with old-world stonemason skills, the results were wonderful.

Because it is relatively easy to work with and its lithology allows for artistic cutting and dressing, Fencepost Limestone has been used as building stone well beyond its native outcrop zone, particularly to the west where limestone and chalk of the Niobrara Formation dominate the surface geology. County courthouses in the towns of Mankato, Osborne, and Hays were constructed from Fencepost Limestone even though the Fort Hays Limestone of the Niobrara is abundantly available locally. Also located in the Fort Hays zone, the handsome Saint Joseph Catholic Church in the village of Damar was built of Fencepost Limestone. To the east, the Lebold Mansion in the Flint Hills town of Abilene, considered one of the "Eight Wonders of Kansas Architecture," was built primarily of Fencepost Limestone.

Rock of the Niobrara Formation underlies the western third of the Smoky Hills and is composed of two main subunits. The Fort Hays Limestone, the older, lower member, crops out in the eastern portion of the Niobrara zone. The Smoky Hill Chalk member crops out to the west. Both have been used as building stone.

Fort Hays Limestone has been used extensively from the vicinity of Hays west through the watersheds of the Smoky Hill, Saline, and Solomon Rivers. Many older buildings in downtown Hays and on the campus of Fort Hays State University were constructed from this locally abundant stone. A number of churches in the region were built with Fort Hays Limestone, including Saint Catherine in Catharine and Saint Mary in Ellis, both imposing Catholic cathedrals, and humble country chapels like Ash Rock Church in Rooks County, Emanuel Lutheran in Trego County, and Saint Andrew's Episcopal in Ellis County. Fort Hays Limestone was also used in many Depression-era projects in the region, including the township hall in the Exoduster settlement of Nicodemus.

Smoky Hills Chalk is less durable than Fort Hays Limestone, and its use as structural stone was fairly localized in the driest parts of western Kansas, like the ranch buildings on The Nature Conservancy's Smoky Valley Ranch. A fair number of commercial buildings and churches in Hill City were built of Smoky Hill Chalk, but use of this stone, which weathers to white, yellow, and a distinctive pinkish-orange color, seems to have peaked in Depression-era projects like the high school football stadium in Oakley, the municipal auditorium (now museum) in Leoti, and the city hall in Colby.

Stone of the Ogallala Formation appears sparingly in the built environment, mostly because it crops out where there isn't much built environment. Exposures of this coarse-textured sandstone, the bedrock of the thinly populated High Plains, typically occur in the landscape as ledges along ravines and streams. You have to do some serious searching to find structures built of this rock, but there is Saint Joseph Catholic Church in New Almelo and an austere but sturdy little chapel located in a remote corner of Hitchcock County, Nebraska, built in 1900 and simply known as the Stone Church.

The most distinctive Ogallala structures I've come across are on the Cottonwood Ranch in Sheridan County, now a state historic site. The ranch buildings, which date to the 1880s, include a house and washhouse constructed of Ogallala blocks artfully dressed at the corners and windows with pink-tinted trim of yellowish Smoky Hill Chalk, both rock types quarried nearby. Immigrants from the Yorkshire region of northern England established this ranch along the South Solomon River, bringing with them hardy Merino sheep and Anglo-Saxon architectural taste and stoneworking skills.

Certain outcrops of Ogallala rock in western Kansas contain irregular nodules of chert, a waxy, resinous, opalized stone. The striking city hall in Bird City, a Depression-era project, is constructed from this unique and highly localized Ogallala variant, as are several structures in the town park. The largest chunks of chert were used as ornament in decorative lintels above the windows, and they also adorn the low rock wall that encircles the building.

Despite its general unsuitability for construction, one zone within the Ogallala outcrop area in western Kansas and a bit of adjacent Nebraska yields serviceable building stone of distinctive character. In this local variant, known in the past as Woodruff granite and Bloomington quartzite, the cementing material is silica-derived opal rather than calcium carbonate, which gives the stone greater density and hardness and a unique greenish hue. Classified as orthoquartzite by geologists, this rock has been used as structural stone in towns in and around its outcrop zone. It has been used most enthusiastically in Hill City, where it imparts a distinctive jade-green tint to the town, especially in the community park with its extensive rock walls and charming bandstand.

During the Great Depression of 1929 to 1939, millions of Americans were out of work and facing desperate economic hardship. Things were even worse in the central and southern Great Plains, where unprecedented drought from 1935 to 1937 caused crop failures and such massive wind erosion of agricultural lands that the region was dubbed the Dust Bowl. President Franklin D. Roosevelt, elected in 1932, promised a "new deal" for the "forgotten man" and launched a host of federally funded projects to create jobs and stabilize the economy.

While national in scope, these federal relief programs were particularly focused on rural areas in Dust Bowl states. Two of these New Deal programs, the Works Progress Administration (WPA) and the Public Works Administration (PWA), provided funding for the construction of civic buildings and public infrastructure improvements and were particularly important in plains states like Colorado, Kansas, and Nebraska.

Structures built with the help of New Deal funding can be found throughout the country of the Kaw, but those constructed of native stone are most common in the rock towns of the Smoky Hills and eastern High Plains. Here, on the outer edges of the Dust Bowl, the availability of structural stone that could be obtained nearby allowed authorities to orchestrate public works projects that would employ large numbers of local, often unskilled, workers while keeping material costs as low as possible. For example, the building of a new community auditorium in the town of Phillipsburg employed about 150 area workers between 1935 and 1937. Stone for the building was quarried from an outcrop of green Ogallala quartzite south of town.

The aesthetics of these New Deal projects varies from town to town. In situations where both high-quality stone and skilled stonemasons were available, the resulting structures were as beautiful and architecturally refined as any built during more prosperous times. The Jewell County Courthouse in Mankato and the Ellis County Courthouse in Hays, both constructed of Fencepost Limestone, are shining examples.

Certain types of rock used in New Deal projects, while locally abundant, were not suited for cutting or dressing, even under the saw and chisel of an expert stonemason. In these cases, the builders often went for a more rough-hewn or "rusticated" look. The city hall in Bird City, constructed of Ogallala stone, and the library (now museum) in Norton, built of green Ogallala quartzite, are prime examples. While not as elegant as similar structures built from Fencepost Limestone, they have a ruggedness and individuality that befits a High Plains sense of place.

In addition to civic buildings, New Deal projects included public infrastructure improvements like waterworks and bridges. Most of these were pretty mundane, but two in Kansas are celebrated yet today for their beauty and artistry. One is the striking stone water tower in Paradise, a silo-like structure thirty-five feet tall that stands like an obelisk at the entrance to town. The other is the Fort Fletcher Stone Arch Bridge, a graceful four-arch bridge spanning Big Creek south of the town of Walker. Both were constructed from locally quarried Fencepost Limestone, and both are on the National Register of Historic Places.

The New Deal also put an architectural stamp on many community parks in the country of the Kaw. Native stone structures often abound in these places, including picnic shelters, gazebos, amphitheaters, bandstands, and swimming pool bathhouses. Many of these are still in use today and are treasured local landmarks, like the Dakota sandstone band shell in Lindsborg's Swensson Park.

These park projects often included a considerable amount of landscape rockwork in the form of entrance gates, walls, bridges, decorative fountains, fireplaces, even grottos. These structures were mostly built in a rustic character from "fieldstone" that could be gathered along creeks and streams in the area, rather than stone that was quarried and cut. McNish Park in Fairbury, Nebraska, is chock-full of structures built by WPA workers in 1935 using Dakota sandstone, so much so you get the feeling the authorities were trying very hard to keep the local guys busy.

Buildings constructed with New Deal support often have a small (seven-by-twelve-inch) bronze plaque mounted near the entrance stating something like "Erected by the Works Progress Administration and the City of Hill City, A-D-1938." But even if you can't find such a marker, there is an architectural signature to many of these projects that gives them away, a design that had its origins in an event that occurred in, of all places, the fancy-schmancy city of Paris.

The *Exposition Internationale des Arts Decoratifs et Industriels Modernes* in 1925 was a world's fair of sorts, devoted to displaying avant-garde approaches to architecture and other decorative arts. The style that emerged from the event, later given the name "art deco," featured clean lines, simple forms, and geometric ornament reflective of the streamlined aesthetic of the machine age.

Art deco architecture was quite different from that which characterized most of the public buildings constructed in the country of the Kaw during the first decade of the 1900s, such as the native stone courthouses of Clay, Douglas, Lincoln, Mitchell, Osborne, and Riley Counties, which were designed in the elaborate, even grandiose, Richardsonian Romanesque style, complete with splendid clock towers. Though stately works of art, these civic emblems probably seemed a bit old-fashioned, even dowdy, after World War I.

The most spectacular art deco structure in the realm of the Kaw is the Kansas City Power and Light Building, built in 1931. This elegant skyscraper, thirty-one stories high and crowned with a lantern-like, ninety-seven-foot-tall finial, was for a time the tallest building in the state of Missouri. Square at the ground floor but tapered through a series of setbacks to create a telescoping, vertical silhouette, the building has been called "a masterpiece of art deco design."

Though not physically situated within the Kaw watershed, the unmistakable Power and Light Building is clearly visible from Kaw Point and Strawberry Hill. So iconic is this building to the KC sense of place that Thomas Hart Benton slipped it into the corner of his painting *Flood Disaster (Homecoming—Kaw Valley)*, which depicted a devastated bottomland neighborhood in the aftermath of the 1951 flood. I have long loved the sight of it, especially since my dad worked there in the 1960s, so I'm claiming it for the country of the Kaw.

Under the influence of tastemakers running New Deal programs in the 1930s, art deco styling was applied to the design of many civic buildings in the Great Plains, but in a manner more suited to small-town sensibilities. The Jewell County Courthouse in Mankato is said to exhibit a "restrained" art deco style, while the architecture of the city hall in Colby is described as "minimalized" art deco. As noted in documentation to place the Colby City Hall on the National Register of Historic Places, "The simple, modern design executed in native stone presented the citizens of Colby with an image that was both progressive and reliable."

So, thanks to a global economic meltdown, an approach to design dreamed up by artists sipping espressos in Parisian cafés left its mark on the country of the Kaw, from the Kansas City skyline out to the headwaters of Prairie Dog Creek.

Glorious native stone courthouses, churches, and other aspirational buildings can be found throughout the country of the Kaw, but what I enjoy most about a rock town reconnaissance is the discovery of an unassuming New Deal gem tucked away on Elm Street. Even the tiniest village might harbor one, like the township hall in Bogue, Kansas, population 250, built in 1936 of green Ogallala quartzite, and the auditorium/gymnasium in Schoenchen, Kansas, population 170, built in 1940 of Fencepost Limestone.

While the native stone structures erected during heady post-frontier boom times proclaim chamber-of-commerce optimism for the future, these Depression-era projects speak of hope in the midst of crippling economic and environmental disaster—local acts of resolve in the face of titanic, global forces.

I find them far more heroic.

Bookend scenes from the country of the Kaw. *Top: Early Evening at Kaw Point* by Craig Thompson, looking east from Kaw Point in Kansas City, Kansas, showing confluence of the Kaw with the Missouri River and the Kansas City, Missouri; the skyline is beyond, the iconic Kansas City Power and Light Building is visible on the right. *Bottom:* Cottonwood grove on the Arikaree River, westernmost tributary of the Kaw, on The Nature Conservancy's Fox Ranch near Idalia, Colorado. Photo by Dave Showalter.

Green by Scott Bean Photography. Konza Prairie Biological Station near Manhattan, Kansas, showing tallgrass prairie and characteristic bench-and-slope cuesta topography of the Flint Hills.

Summer Afternoon Views by Scott Bean Photography. Image of Smoky Hill Buttes taken from Coronado Heights near Lindsborg, Kansas, showing characteristic Dakota Hills topography of the eastern Smoky Hills.

Post Rock Country landscape in the middle Smoky Hills showing mixed-grass prairie rangeland and posts quarried from the Fencepost Limestone bed of the Greenhorn Limestone formation. Photo by Harland J. Schuster.

Little Jerusalem Badlands State Park in the western Smoky Hills showing eroded Smoky Hill Chalk of the Niobrara Formation with dead-level High Plains surface on the horizon. Photo by Bruce L. Hogle.

Aerial view of High Plains landscape with playa basins, Scott County, Kansas. Photo by William C. Johnson, courtesy Kansas Geological Survey.

Arikaree Breaks, Cheyenne County, Kansas, showing High Plains aeolian (wind-shaped) landscape of deeply eroded loess deposits with plains yucca plants in bloom in the foreground. Photo by Harland J. Schuster.

Floral tributes to the Great Pathfinder. Fremont's clematis (*top*) and Fremont's evening primrose (*bottom*), signature wildflowers of the Smoky Hills, were named in honor of John Charles Frémont, who collected specimens of each in 1844 as his exploring party returned to "the little town of Kansas" from an expedition to the Pacific Coast. Photos by Jim Locklear.

Scenes from the Rainwater Basin. *Top:* Kirkpatrick Basin North Wildlife Management Area near York, Nebraska. Photo by Eric Fowler, courtesy Nebraskaland Magazine/Nebraska Game and Parks Commission. *Bottom:* Franklin's gulls at rest in a wetland near Utica, Nebraska, showing pinkish blush of spring breeding plumage due to a diet of marine crustaceans consumed while the birds wintered on the Pacific coast of South America. Photo by Joel G. Jorgensen.

Waters of the Kaw. *Top:* Kaw River near Lecompton, Kansas, showing relatively unaltered riverscape with islands, sandbars, shallow side channels, and backwaters, the Delaware River entering from the north. Photo by Craig Thompson. *Bottom:* Minnows in Illinois Creek, a pristine Flint Hills stream. Shown are southern redbelly dace (black side stripe, red belly), carmine shiner (rose-colored head, silver sides), and common shiner (largest). This remarkable photo was taken in-stream by Michael Forsberg (www.michaelforsberg.com).

Native stone gems of the Great Depression. The Jewell County Courthouse in Mankato, Kansas (*top*) and the City Hall of Bird City, Kansas (*bottom*) were each constructed in 1936 under the auspices of the federal Works Progress Administration, the courthouse with Fencepost Limestone and the city hall with a local variant of the Ogallala Formation. Both reflect restrained art deco architectural styling. Photos by Jim Locklear.

TWENTY

Stickers

> The universities now offer only one serious major: upward mobility. Little attention is paid to educating the young to return home, or to go some other place and dig in. There is no such thing as a "homecoming" major.
> Wes Jackson, *Becoming Native to this Place*

Frederick Albertson was not a rambling man. Born in 1892 in Hill City, Kansas, he attended school in a sod schoolhouse in rural Graham County. In those days, the education of a western Kansas farm boy was considered complete when he finished eighth grade. But Albertson kept going, earning a bachelor's degree from Fort Hays Normal School (later Fort Hays State University) in 1918 and eventually a PhD in botany from the University of Nebraska under the direction of Professor John E. Weaver. Albertson spent his entire academic career at Fort Hays State, about forty miles from his Smoky Hills birthplace, becoming professor of botany in 1937. Like his mentor Weaver, Albertson became a world-renowned grassland ecologist, with unparalleled firsthand knowledge of the mixed-grass prairie. He passed away in 1961 while still on the faculty.

Fred Albertson was a sticker.

The notion of *boomers* and *stickers* comes to us from the writings of Wallace Stegner, acclaimed chronicler of life in the American West. It was Stegner's observation that two types of people are drawn to a new country such as the West once was. There are the boomers, those who "pillage and run," who approach the land and its resources "as grave robbers might approach the tomb of a pharaoh." And there are the stickers, men and women who put down roots and "love the life they have made in the place they have made it."

There are plenty of boomers in the environmental history of the country of the Kaw. Now we'll get acquainted with a few of the stickers.

Not long ago, I visited the Kansas State Capitol in Topeka to see Pete Felten's celebrated sculptures, *Four Famous Kansans*. The imposing figures, each about eight feet tall and carved from native Kansas limestone, depict Amelia Earhart, Dwight Eisenhower, William Allen White, and Arthur Capper.

The sculptures are arrayed around the inside perimeter of the capitol rotunda, as were that day a smattering of lobbyists, dressed to kill and yakking on cell phones or talking to one another in hushed tones, all the while furtively glancing around to see who was coming and going. Depending on the legislation they had been hired to influence, one person they probably were hoping to *not* see was Dawn Buehler.

Dawn Buehler is the executive director of Friends of the Kaw, a nonprofit with a simple but mighty mission: "To protect and preserve the Kansas River." The organization was founded in 1991, and Dawn became the third executive director in 2015. Because Friends of the Kaw is an affiliate of the Waterkeeper Alliance, an international network of grassroots organizations dedicated to "drinkable, fishable, swimmable waters," Dawn also holds the title of Kansas Riverkeeper.

The focus of the Friends is the 173-mile mainstem of the Kaw, from the junction of the Republican and Smoky Hill Rivers to confluence with the Missouri. The waters of this stretch, designated the Kansas River National Water Trail in 2012, are public and open for use without the permission of adjacent landowners. Dawn speaks of the Kaw as "the people's river," and is proud that the Friends have helped establish nineteen boat ramps along its length, facilitating access and, in a very real sense, giving the Kaw back to the citizens of the state.

As the self-described "eyes, ears, and voice of the Kaw," Dawn is not shy about buttonholing politicians or going toe to toe with special interest groups. She is a familiar figure at the statehouse and has served on numerous state advisory boards, most recently as the governor-appointed chair of the Kansas Water Authority.

But this daughter of the Kaw prefers to get her work done on the river.

A sixth-generation Kansan, Dawn's passion is rooted in a childhood spent on the family farm situated along the Kaw near the town of De Soto, about halfway between Kansas City and Lawrence. Aptly named Riverview Farms, the two thousand acres of rich floodplain soil produced fine crops of potatoes, sweet potatoes, sweet corn, tomatoes, watermelon, turnips, and other vegetables, most of which were sold to a chain of grocery stores. After high school, Dawn helped on the farm while going to college (the first in her family to do so). A catastrophic flood in 1993 caused her family to cease farming in the Kaw valley.

Dawn grew up fishing on the Kaw with her dad; they boated on the river and camped on its sandbars long before these became popular pastimes. Her bond with the Kaw was deepened by watching sunsets over the water from the De Soto bridge. She has since paddled every one of the Kaw's 173 water miles. For Dawn, the well-being of this particular river is a particularly personal matter.

The Friends of the Kaw is a very hands-on conservation and environmental organization. Each year the group orchestrates a number of trash cleanups, some of these efforts focused along the banks and others in the stream. Old tires are particularly abundant in the Kaw channel; used in years past for bank stabilization, they were later washed out by floods and carried downstream. Plastic battery casings, the other big source of refuse, are thought to have entered the river from a now-defunct recycling plant. Much of this trash ends up embedded in sand bars. Strenuous effort is required to extract and haul it to the riverbank, and it takes lots of volunteer labor plus close coordination with local authorities to dispose of the trash.

The Friends also take on other forms of environmental degradation. The biggest battles have been over dredging—the mining of sand and gravel from the river channel. Dredging operations result in channel widening, bank erosion, riverbed degradation, and altered stream velocities, all of which affect the distribution and abundance of fish, freshwater mussels, and other aquatic life. Because sand and gravel are essential materials for construction, the Friends are working to shift commercial extraction toward privately owned floodplain properties and away from the sandbars and riverbed of the main channel, which is public land.

But Dawn feels the most effective work of the organization happens on their group paddle trips. In her experience, nothing stirs affection and sympathy for the river quite like floating along it in a canoe or kayak, and Dawn uses time on the water to educate and build relationships. One of her paddle trip partners was a Kansas governor not held in high regard by the environmental community. Dawn took a lot of heat on social media for even being seen with him, but considered it a small price to pay for the opportunity to influence the state's water policy.

I arranged to meet Dawn at a coffeehouse in Lawrence so I could learn about the organization she leads plus hear her personal story. We had a lot to talk about, and morning coffee segued into lunch. I'm glad we met in person, because I got to witness the emotion. Dawn Buehler is what you might call a force of nature—passionate, focused, fearless. Yet she teared up twice during our visit, first when reminiscing about her dad, later when talking about her yet-to-be-born first grandchild. Dawn's devotion to the Kaw is anchored in her family history, her advocacy animated by hope of healthier waters for the next generation.

When Thomas Jefferson was putting the team together for what would later be known as the Lewis and Clark Expedition, he wanted to make sure someone would document the plant life of America's newly acquired eight hundred thousand square miles of territory. The task was so important to the horticulturally inclined president that he laid the responsibility on the shoulders of Meriwether Lewis, who as co-commander of the Corps of Discovery already had a lot on his plate. No matter that Lewis didn't know much about botany: Jefferson would just send him to Philly.

In Jefferson's day, Philadelphia was the center of botanical science in America, so the president sent Lewis there in early 1803 to learn how to collect and preserve plant specimens that could later be studied by the scientific community.

No disrespect, Philadelphia, but should the current commander-in-chief have need of botanical wisdom and ask me for advice, I would point to Aurora, Nebraska.

Aurora is a town of forty-six hundred people situated in the east-central part of the state. Lincoln Creek, tributary of the Big Blue River, tributary of the Kaw, runs through the north side of town. Aurora, which has one of the most charming courthouse squares of any town in the plains, is home to one of the most unique conservation organizations in the country of the Kaw—the Prairie Plains Resource Institute.

Bill and Jan Whitney were children of the 1970s. Both graduated from high school in the early part of the decade, Bill from Aurora and Jan from Sandy Creek, a rural school in the next county to the south. The two met in college in Lincoln and were drawn to the countercultural, back-to-the-land ethos of the day. Following marriage and college (and grad school for Bill), family members encouraged them to head somewhere interesting. But with Bill's father battling a serious illness, they landed in Aurora in 1977 and began looking for a meaningful way to make a living and a difference.

Happily, they found work codirecting a prairie interpretation project at a history museum in a nearby town, which tapped into Bill's education in ecology and their mutual interest in connecting people with nature. Wanting to create a more authentic representation of the tallgrass prairie on the museum grounds, they visited trailblazing prairie restoration efforts in Illinois and Wisconsin and returned home inspired and with a sense of calling: they would become artisans in the craft of prairie restoration and perhaps even recreate wilderness in Nebraska corn country.

They would also start an organization devoted to prairie conservation. The Prairie Plains Resource Institute was incorporated as a nonprofit in 1980, but it would be years before it brought in any kind of income. So, the Whitneys worked a variety of jobs during the early years to pay the bills and sustain them and their two daughters, all the while trying to get their fledgling organization off the ground, Bill experimenting with prairie restoration while Jan focused on outreach.

To launch an environmental organization in a small town in a politically conservative county almost entirely given over to corn and soybeans required vision, gumption, and periodic doses of encouragement. Bill and Jan found affirming souls in a remarkable community of artists, writers, and thinkers with ties to the area. These included Ernie Oschner, a sky-loving painter and photographer in Aurora, Norris Alfred, Pulitzer prize–nominated editor of the *Polk Progress* newspaper, and Emiel Christensen, an architect based in Columbus. They also found solace at a place with the suspicious name of the Art Farm, a bohemian hangout back in the seventies but today a legit, nonprofit artist residency organization that draws creative types from across the United States and around the world for an immersion experience in an agricultural landscape.

The work of prairie restoration boils down to two basics—seed and land. Bill's search for seed led to his discovery of remnant stands of prairie around Aurora, places like roadside ditches, railroad rights-of-way, and small native hay meadows. His first harvest consisted of a coffee can full of seeds from thirty-five different species of grasses and wildflowers, and the site of his first planting experiment was a tiny plot of ground at the Art Farm. The next year, 1980, he would establish the first official PPRI planting, sowing seed of close to a hundred different species on some marginal land along Lincoln Creek, not far from the Whitney's home.

Forty years after that start on Lincoln Creek, PPRI has now restored over fourteen thousand acres (twenty-two square miles!) of prairie and wetlands on 323 sites in sixty Nebraska counties. And now, PPRI annually harvests seed from over 225 species. The aim is to create high-diversity plantings that resemble as closely as possible the species richness of remnant, benchmark plant communities in the area. To ensure adaptability to the local ecological conditions, they typically only use seed collected within a 100-mile radius of the restoration site. "High-diversity, local-ecotype" are watchwords for PPRI restoration projects.

Over the course of four decades and hundreds of projects, Bill and the staff at PPRI have developed a host of innovations and techniques to facilitate the hard work of prairie restoration. Combining ecological knowledge with handyman skills and agrarian ingenuity, they have developed or modified all kinds of tools and equipment to make seed harvest, planting, and after-planting management more efficient and effective. The education and outreach side of the PPRI enterprise likewise flourished under Jan's creative attention, resulting in highly regarded publications and innovative programs like SOAR (Summer Orientation About Rivers), a nature day camp on the nearby Platte River.

As for the Philadelphia parallel, the presence of PPRI in Aurora has made the town a magnet and hub for conservation activity in Nebraska. The Nature Conservancy set up an office downtown, as did the state botanist of the Nebraska Game and Parks Commission. It doesn't hurt that

Aurora is fairly close to the Platte River staging area of migrating sandhill cranes, and the shorebird-loved wetlands of the Rainwater Basin, both high-priority targets for habitat restoration efforts, but other larger cities in the region could have been chosen for these offices. PPRI recently opened the Charles L. Whitney Educational Center on their Gjerloff Prairie property north of Aurora, which provides an outstanding setting for prairie restoration workshops.

Bill and Jan retired in December 2019 after forty years of dedication to the Prairie Plains Resource Institute. They still live in Aurora and are recalibrating their lives to the roles of founders rather than managers of the organization they poured themselves into.

Powerful affirmation of the Whitney's work is provided by the recent book *Hidden Prairie* by Chris Helzer. Chris, who lives in Aurora, is the director of science for The Nature Conservancy in Nebraska and an accomplished photographer and author. Beginning in January 2018, he spent a year photographing a single square meter of prairie to document the biodiversity supported by a tiny fraction of this once horizon-sweeping ecosystem. By the end of the year he had photographed 113 different species of plants and animals in the plot, with most of the diversity represented by bees, beetles, and other hard-to-see critters. For the setting of his project, Chris chose a strip of prairie along Lincoln Creek, restored by Bill on land from which soil had been scraped for a construction project.

Few people have more intimate knowledge of the tallgrass prairie than Bill Whitney, not just how the prairie works, but how to put it back together when it is broken. The amazing biological diversity sponsored by that little patch of ground on Lincoln Creek demonstrates the healing capacity of the craft of ecological restoration.

While restoring prairie is at the heart of the mission of the Prairie Plains Resource Institute, the work has been carried out with the conviction that long-term progress in solving environmental problems depends on addressing social and cultural issues as well as technical ones. The aim has been to develop a version of restoration that is adapted to the local culture, married to concepts surrounding small-town economic development, and dedicated to long-term improvements. This unique approach to environmental stewardship is captured in the grand vision of PPRI: "Ribbons of Prairie."

As Bill explained to me recently, this vision is nothing short of the remaking of prairie watersheds from the top down. It is an idea best imagined from the sky, a continuous network of restored prairie wilderness along stream and river corridors, with ribbons of green prairie and blue waters woven into a tapestry of farmlands and communities urban and rural. Restoration in every sense of the word.

Ambitious? That is an understatement. Bill concedes Ribbons of Prairie is a multigenerational, hundred-year aspiration. But the world needs big ideas like this and idealistic doers like the Whitneys. Just imagine, a prairie restoration along Lincoln Creek, in the far reaches of the country of the Kaw, making life richer for the residents of Aurora while also sending cleaner water on to folks living in the Rosedale neighborhood of Kansas City, Kansas.

I, for one, am thankful Bill and Jan Whitney followed their hippie hearts back to Aurora.

Nathan Andrews tends some of the most pristine habitat in the country of the Kaw. He and his wife, Laura, and their young family live on Fox Ranch, a Nature Conservancy preserve in Yuma County, Colorado. The preserve consists of 14,100 acres of rangeland, about 80 percent of it in sandsage prairie, and an eight-mile stretch of the Arikaree River, the westernmost stream in the Kaw watershed.

The Nature Conservancy acquired the Fox Ranch in 2000, and Nathan has been the lessee and manager since 2006. I had been on Fox Ranch a couple of times doing research related to sandsage prairie and had visited briefly with Nathan on both occasions. But I wanted to get a fuller understanding of the relationship between this family and the preserve, so I made the six-hour drive out from Lincoln and shared coffee and cookies with Nathan and Laura at their dining room table.

If you're not familiar with The Nature Conservancy, it is a private nonprofit organization with a mission "to conserve the lands and waters upon which all life depends." Founded in the United States in 1951, the Conservancy is now a global enterprise and works in over seventy countries and territories. A key part of the organization's strategy over the years has been to conserve the "last great places"—purchasing and managing the most ecologically unique yet environmentally vulnerable landscapes remaining on Earth.

Why would the Conservancy be interested in Fox Ranch? The first and foremost reason is the presence of the Arikaree River. The Arikaree (the Andrews and other longtime residents prefer *Arickaree*) represents the last best example of a free-flowing High Plains stream. Today, many of the streams that arise on the High Plains carry little or no water from their headwaters down to their middle reaches, the consequence of center pivot irrigations systems that have lowered the Ogallala aquifer and "decoupled" the streams from the groundwater necessary to maintain streamflow in this semiarid environment.

The Arikaree has not been immune to dewatering. The river has lost about thirty miles of water between its headwaters and Fox Ranch, meaning no water flows in its streambed. The current point of emergence, where water again appears in the Arikaree, is less than a mile above the western boundary of Fox Ranch. Because the stretch of the Arikaree on Fox Ranch almost always carries water, it supports a unique and imperiled community of Great Plains fishes plus a stand of cottonwood and tallgrass riparian vegetation that has all but disappeared from other stream systems in the High Plains.

One of the other major attributes of Fox Ranch is an expanse of sandsage prairie that occupies a large dune field to the north of the Arikaree valley. This community supports healthy populations of a number of grassland birds, including the imperiled greater prairie chicken.

But of greater consequence than the biodiversity supported by the sandsage prairie is the fact that the sandy soil and choppy terrain make much of this area unsuited to farming, which makes it unsuited to irrigation. As a result, the dune field provides a bit of an irrigation buffer zone to the north of the Arikaree, reducing, at least locally, aquifer drawdown. And, as reported in early geological studies of the region, the water-absorbing sand dunes allow a measure of recharge to the Ogallala and alluvial aquifer that does not occur in stretches of the river bordered by flats of shortgrass prairie-turned-cropland. These unique and fortunate circumstances appear to translate into a bit more water in the Arikaree as it passes through Fox Ranch.

In partnership with The Nature Conservancy, Nathan manages Fox Ranch not by treating it as untouchable wilderness, but by working it as grazing land. He is a student of Holistic Management, a ranch management framework originally formulated by Zimbabwean scientist Allan Savory that integrates financial, grazing, infrastructure, and land management elements into a strategy that balances sustainability with business goals. The grazing part of this strategy incorporates intensive, short-duration grazing rotations of livestock through multiple pastures or paddocks. This strategy is not without controversy in the range management profession, but Nathan and Conservancy scientists have seen it achieve positive results on Fox Ranch.

Spreading out a large map of the ranch on the dining room table, Nathan shows how the twenty-three square miles of property are divided into multiple pastures, each delineated by a run of electric fence. It looks like a giant puzzle made up of a bunch of randomly shaped polygons, but the boundaries of each pasture are determined by in-depth knowledge of the topography of the land and differences in the species makeup of the vegetation. Decisions about how long to graze a particular pasture and when to move cattle to another require intimate familiarity with each pasture and its forage production capacity, plus continuous monitoring of current range conditions, particularly during times of drought.

Using Black Angus as his tool, Nathan has been able to shape the structure of vegetation on the ranch, often at a surprisingly fine level of detail. For example, greater prairie chickens prefer a particular structure to the vegetation, especially during nesting and brood rearing, and Nathan uses rotational grazing to achieve a target of fourteen inches of grass cover on 30 percent of the ranch. Such fine-tuning also improves cattle production on the land. Sand bluestem is a dominant grass in healthy stands of sandsage prairie, but, because cattle love it like candy, it is often suppressed or even grazed out of rangeland. Nathan has seen this important forage species increase at Fox Ranch under his carefully managed grazing strategy.

I've studied a lot of grassland over the years but know absolutely nothing about managing a ranch or running cattle. Yet it seems to me that the approach to range management practiced by Nathan borders on art, requiring deep knowledge of both grass and cows, love of the land, and pride, even joy, in achieving good stewardship. It also seems to me that Nathan was born for this line of work.

The Andrews are not newcomers to the Arikaree "neighborhood." Nathan grew up to the east of Fox Ranch, not far from the town of Idalia (current population of thirty-seven), where his folks met in school. He is the sixth generation on the land on his dad's side of the family and fifth on his mom's side. Laura is third generation in the area, her people being from the vicinity of Arriba and Anton, villages to the west of Fox Ranch.

Although only thirty-eight years old, Nathan seems like a bit of an old soul as he speaks of the ethic of stewardship he learned from his father and grandfather and his desire to live out his years in Arikaree country, land some might find austere but which he and Laura clearly love. He also wants to help the next generation find a way to make a living on the land.

As I was wrapping up the visit and saying my goodbyes, Nathan asked the kids if there was anything they thought "Mr. Jim" should know about the ranch. The oldest, their seven-year-old daughter, spoke up and told me that "Boss" had died the day before. I wasn't sure what to say to this until I realized that Boss was the family dog, and not a relative or friend. They pointed out his little bed, still sitting next to the dryer in the laundry area, which the children had decorated with drawings to comfort their beloved border collie during the last days of his struggle with cancer.

The tender care extended to Boss is emblematic of the devotion with which this family is attending to Fox Ranch. Let's hope the Andrews and folks like them can continue to find a place to live and love in this far-flung corner of the country of the Kaw.

For their sake and ours.

In 1949, as outgoing president of the Kansas Academy of Science, Frederick Albertson gave a farewell address to his colleagues titled, "Man's Disorder of Nature's Design in the Great Plains." In it he decried the alarming rate at which grassland was being overgrazed or completely plowed under in western Kansas and eastern Colorado. He concluded his talk with the following admonition: "It should be the policy of all who live and work in the plains region to learn more of its proper use and, at the same time, how to preserve its beauty."

The great ecologist's call to action was in essence an admonition to live like true plainswomen and plainsmen, and it describes what Dawn Buehler, Bill and Jan Whitney, and Nathan and Laura Andrews have given themselves to. Thankfully, they are not the only ones. A whole host of folks across the country of the Kaw are working on their own or through organizations, associations, and agencies for the sake of conservation, preservation, restoration, and renewal. Some have lived here all their lives, others came from someplace else and, in the words of Wes Jackson, decided to "dig in."

They're all stickers, in my book.

Epilogue

> This looking business is risky.
> *Annie Dillard,* Pilgrim at Tinker Creek

The instant I realized what I was seeing, I wanted desperately to *unsee* it. I even looked away for a moment, to prevent my eyes from taking in more imagery.

It was early October and I was out in the western reaches of the Rainwater Basin. I was driving along a gravel road south of the village of Atlanta, Nebraska, when out of the corner of my eye I glimpsed a pheasant-sized bird sitting on the ground in a recently harvested corn field. I stopped to take a closer look and was surprised to see it was a Swainson's hawk.

I thought it odd for a hawk to be sitting in corn stubble and wondered if it had been injured. Then I noticed all the other Swainson's scattered across the field. A couple dozen of them. Some were just sitting there while others were hopping across the ground like oversized robins, no doubt in pursuit of grasshoppers exposed by the harvesting. It struck me as highly undignified behavior for a raptor, especially the signature summer hawk of the wide-open plains.

Then I looked up, and my heart sank. The sky was filled with kettling Swainson's.

You may recall that this sight, a spiraling kettle of Argentina-bound Swainson's hawks, circling ever higher to catch thermals, was an item on my bucket list, something I dreamed of experiencing before I died.

But not like this.

I fancied that my first sight of kettling Swainson's would be in the skies above a pristine sweep of shortgrass prairie, maybe with a small band of pronghorn skimming over the horizon. But, instead of wilderness, the tableau would be a corn field. I really did consider looking away and driving off, so that my first kettle wouldn't be this kettle. But it was too late.

As if to spoil my experience even more, off in the distance I could see the stark, sixty-five-foot-tall chimney that marks the site of Camp Atlanta, a World War II internment facility for German prisoners of war. The story of Camp Atlanta is actually rather heartwarming: the prisoners worked on farms in the area and a few became so fond of the place and people that they returned after the war and became US citizens. The sprawling complex was completely dismantled in 1947 except for a water tower and the imposing power plant chimney.

But on this day, in my increasingly dark mood, the sight of Camp Atlanta reminded me of images of Nazi death camps, where the most recognizable structures are the unspeakable chimneys. Despite the benign environment of Camp Atlanta, it existed because of a war precipitated by a worldview that justified the extermination of human beings, infants to aged, on an industrial scale.

This was not how I dreamed of experiencing a kettle of Swainson's hawks. And this is not how I thought I would wrap up this book.

I was hoping to find some grand moment or poignant story to serve as the climax and awaken affection and sympathy for the country of the Kaw. Kettling Swainson's would have been the perfect literary vehicle for opening and closing this project, but those cornfield-scrounging birds ruined everything.

Or did they?

I've visited a lot of small towns and villages throughout the country of the Kaw, close to three hundred over the years, and I do have my favorites. Sometimes I'm taken in by the loveliness of the setting—Akron riding the prow of the Colorado High Plains, Stockville in the breaks of Nebraska's Medicine Creek country, Onaga in that vale in the Kansas Flint Hills. And I'm sure time of day has something to do with it, with more admiration sparked by an encounter at dusk or dawn. But usually there is something about the personality of the place—an unaffected peculiarity—that tugs at my heart.

The little village of Dwight, Nebraska, is such a town. Home to 188 souls, Dwight is situated on the divide between the watershed of Plum Creek, tributary of the Big Blue River, tributary of the Kaw, and that of Oak Creek, tributary of Salt Creek, tributary of the Platte. To the west of Dwight are the loess plains of the Rainwater Basin; to the east a hilly region known, "with a wink," as the Bohemian Alps.

This rumpled country is the setting for *Local Wonders*, a delightful little book by Ted Kooser, resident of the Bohemian Alps and, by the by, thirteenth poet laureate of the United States. The regional name is a good-humored nod to the hilly terrain, pretty strong for this part of Nebraska, and the prevalence of Czech communities scattered throughout—Dwight among them, along with Abie, Bee, Bruno, Garland, Prague, and Touhy.

Kooser, who also has a Pulitzer Prize to his credit, lives with his wife, Kathleen, out in the hills and for many years kept a one-room writing studio in downtown Dwight with "Poetry Made and Repaired" lettered on the storefront window. *Local Wonders* is a collection of stories and reflections about the unspectacular but marvelous happenings that distinguish a place to someone who keeps an eye out for them.

Aside from Kooser's studio and a couple of taverns, the most significant building stock in Dwight is associated with the Assumption of the Blessed Virgin Mary Catholic Church. The church itself is a handsome structure of yellow-blond brick, and its scale nicely fits the size of the town. Associated with the church is a cluster of smaller structures plus a sculpture garden that includes two stone grottos. Inside one grotto is a scene depicting Jesus praying in the Garden of Gethsemane. In the other, a sculpture of Saint Jude Thaddeus.

Not being familiar with Roman Catholic saints, I was glad to find a small marker in the shrine with the outline of a suggested prayer. In the gloom of the grotto, I read that Jude is the patron saint of "hopeless cases and of things despaired of." Suddenly, a light switched on. Here was a heavenly advocate for the country of the Kaw! The patron saint of damaged goods.

You can sometimes get a nice deal on a washing machine if you're willing to settle for an appliance that has a few dents or scratches from careless handling. The outside may be a bit dinged up, but it runs just fine. Damaged goods.

But most times we use this expression we're referring to a person, almost always a woman, whose appearance and well-being have been diminished by mistreatment or even abuse. And it is typically uttered in a whisper, sometimes in pity, sometimes in contempt: "She's damaged goods." A ruthless man might even try to take advantage of her vulnerability.

Yet, as we know from the storyline of many a romance movie, true love looks past the wounded exterior and spends itself in binding up the brokenness, restoring both wholeness and beauty, and protecting from future assaults.

The population of Dwight peaked in 1930 at 323 residents. Like hundreds of towns and villages scattered across the Great Plains, Dwight is a shadow of its former self, and that shadow is growing dimmer every year. Assaulted by the vagaries of international trade deals, changing ag technologies, and a welter of other socioeconomic forces, Dwight could use a little help from Saint Jude.

Just a few miles from Dwight is a Rainwater Basin wetland where I have seen Hudsonian godwits in May, recently arrived from the tip of South America and refueling so they can complete the rest of their mind-boggling journey north to the Arctic hinterlands. Like the Swainson's hawks spiraling over that corn field near Atlanta, Hudsonian godwits have somehow adapted themselves and their wondrous way of life to a profoundly broken landscape. We should take inspiration from such resiliency.

My hope for the country of the Kaw is that it will be a land where Dwights and godwits can flourish together. Just like in the movies, it will require true love, costly love.

Appendix A: Plants Mentioned in the Text

Trees

American beech	*Fagus grandifolia*
American chestnut	*Castanea dentata*
American elm	*Ulmus americana*
American sycamore	*Platanus occidentalis*
Aspen	*Populus tremuloides*
Basswood	*Tilia americana*
Bitternut hickory	*Carya cordiformis*
Blackjack oak	*Quercus marilandica*
Black oak	*Quercus velutina*
Black walnut	*Juglans nigra*
Bur oak	*Quercus macrocarpa*
Chestnut oak	*Quercus muhlenbergii*
Dwarf chestnut oak	*Quercus prinoides*
Eastern cottonwood	*Populus deltoides*
Eastern hemlock	*Tsuga canadensis*
Eastern red cedar	*Juniperus virginiana*
Green ash	*Fraxinus pennsylvanica*
Hackberry	*Celtis occidentalis*
Ironwood	*Ostrya virginiana*
Kentucky coffeetree	*Gymnocladus dioicus*
Osage-orange	*Maclura pomifera*
Pawpaw	*Asimina triloba*
Persimmon	*Diospyros virginiana*

Pin oak	*Quercus palustris*
Ponderosa pine	*Pinus ponderosa*
Post oak	*Quercus stellata*
Redbud	*Cercis canadensis*
Red oak	*Quercus rubra*
Rocky Mountain juniper	*Juniperus scopulorum*
Russian olive	*Elaeagnus angustifolia*
Sassafras	*Sassafras albidum*
Shagbark hickory	*Carya ovata*
Shingle oak	*Quercus imbricaria*
Shumard's oak	*Quercus shumardiana*
Siberian elm	*Ulmus pumila*
Silver maple	*Acer saccharinum*
Sugar maple	*Acer saccharum*
Swamp white oak	*Quercus bicolor*
Tulip poplar	*Liriodendron tulipifera*
White ash	*Fraxinus americana*
White oak	*Quercus alba*

Grasses and other graminoids (grass-like plants)

Big bluestem	*Andropodon gerardii*
Blue grama	*Bouteloua gracilis*
Blowout grass	*Redfieldia flexuosa*
Buffalograss	*Buchloe dactyloides*
Hairy grama	*Boutleloua hirsuta*
Indian grass	*Panicum virgatum*
Intermediate wheatgrass	*Thinopyrum intermedium*
Little bluestem	*Schizachyrium scoparium*
Needle-and-thread	*Hesperostipa comata*
Prairie dropseed	*Sporobolis heterolepis*
Prairie sandreed	*Calamovilfa longifolia*
Sand bluestem	*Andropogon hallii*
Sand dropseed	*Sporobolis cryptandrus*
Sandhills muhly	*Muhlenbergia pungens*
Sideoats grama	*Bouteloua curtipendula*

Sun sedge	*Carex heliophila*
Switchgrass	*Sorghastrum nutans*
Threadleaf sedge	*Carex filifolia*

Shrubs

Big sagebrush	*Artemisia tridentata*
Sand sagebrush	*Artemisia filifolia*

Wildflowers (woodland herbs and prairie forbs)

Bitterweed	*Tetraneuris scaposa*
Bloodroot	*Sanguinaria canadensis*
Blue wild-indigo	*Baptisia australis*
Bracted spiderwort	*Tradescantia bracteata*
Bractless mentzelia	*Mentzelia nuda*
Bush morning-glory	*Ipomoea leptophylla*
Butterfly milkweed	*Asclepias tuberosa*
Downy yellow violet	*Viola pubescens*
Dutchman's breeches	*Dicentra cucullaria*
Dwarf milkweed	*Asclepias uncialis*
Easter daisy	*Townsendia exscapa*
Fawn-lily	*Erythronium albidum*
Fremont's clematis	*Clematis fremontii*
Fremont's evening primrose	*Oenothera macrocarpa* subsp. *fremontii*
Fringed puccoon	*Lithospermum incisum*
Golden Alexanders	*Zizia aurea*
Ground plum	*Astragalus crassicarpus*
Hedge bindweed	*Calystegia sepium*
Hoary puccoon	*Lithospermum canescens*
Hooker's sandwort	*Eremogone hookeri*
Jack-in-the-pulpit	*Arisaema triphyllum*
Kansas antelope sage	*Eriogonum jamesii* var. *simplex*
Lavender-leaf evening primrose	*Oenothera lavandulifolia*
Lemon scurfpea	*Psoralidium lanceolatum*

Mayapple	*Podophyllum peltatum*
Mead's milkweed	*Asclepias meadii*
Milkwort	*Polygala alba*
Missouri evening primrose	*Oenothera macrocarpa* subsp. *macrocarpa*
Moss campion	*Silene acaulis*
Narrow-leaf bluet	*Houstonia nigricans*
Narrow-leaf purple coneflower	*Echinacea angustifolia*
New Jersey tea	*Ceanothus americanus*
Nuttall's violet	*Viola nuttallii*
Othake	*Palofoxia sphacelata*
Oval-leaf bladderpod	*Physaria ovalifolia*
Pale purple coneflower	*Echinacea pallida*
Pink poppy mallow	*Callirhoe alcaeoides*
Plains penstemon	*Penstemon albidus*
Plains phlox	*Phlox andicola*
Plains skullcap	*Scutellaria resinosa*
Plains wild-indigo	*Baptisia bracteata* var. *leucophaea*
Plains yucca	*Yucca glauca*
Prairie larkspur	*Delphinium virescens*
Prairie parsley	*Polytaenia nuttallii*
Prairie phlox	*Phlox pilosa*
Prairie ragwort	*Packera plattensis*
Prairie spiderwort	*Tradescantia occidentalis*
Prairie violet	*Viola pedatifida*
Purple milkweed	*Asclepias purpurascens*
Sandhills fleabane	*Erigeron bellidiastrum*
Sand lily	*Leucocrinum montanum*
Sand milkweed	*Asclepias arenaria*
Scarlet globe-mallow	*Sphaeralcea coccinea*
Showy gilia	*Ipomopsis longiflora*
Silky milkvetch	*Astragalus sericoleucus*
Silky prairie clover	*Dalea villosa*
Silver-mounded cryptantha	*Oreocarya cana*
Small-flowered gaura	*Oenothera curtiflora*
Solomon's seal	*Polygonatum biflorum*
Spider milkweed	*Asclepias viridis*

Spreading nailwort	*Paronychia jamesii* var. *depressa*
Spring beauty	*Claytonia virginica*
Stemless four-nerve daisy	*Tetraneuris acaulis*
Summer milkvetch	*Astragalus hyalinus*
Texas sandwort	*Minuartia michauxii* var. *texana*
Timber phlox	*Phlox divaricata*
Toothwort	*Cardamine concatenata*
Topeka purple coneflower	*Echinacea atrorubens*
Wake-robin	*Trillium sessile*

Appendix B:
Animals Mentioned in the Text

Crustaceans

Clam shrimp	Class Branchiopoda; orders Spinicaudata and Laevicaudata
Fairy shrimp	Class Branchiopoda; order Anostraca
Tadpole shrimp	Class Branchiopoda, order Notostraca; family Triopsidae

Insects

Evening primrose sweat bee	*Lasioglossum (Sphecodogastra) oenotherae*
High Plains grasshopper	*Dissosteira longipennis*
Regal fritillary	*Speyeria idalia*
Rough harvester ant	*Pogonomyrmex rugosus*
Scott riffle beetle	*Optioservus phaeus*
Two-spotted bumblebee	*Bombus bimaculatus*
White-lined sphinx	*Hyles lineata*
Yucca moth	*Tegeticulla yuccasella*

Fish

Blackside darter	*Percina maculata*
Blue catfish	*Ictalurus furcatus*
Bluntnose minnow	*Pimephales notatus*
Brassy minnow	*Hybognathus hankinsoni*

Carmine shiner	*Notropis percobromus*
Channel catfish	*Ictalurus punctatus*
Common shiner	*Luxilis cornutus*
Creek chub	*Semotilus atromaculatus*
Fathead minnow	*Pimephales promelas*
Flathead catfish	*Pylodictus olivaris*
Flathead chub	*Platygobio gracilis*
Johnny darter	*Etheostoma nigrum*
Northern plains killifish	*Fundulus kansae*
Paddlefish	*Polyodon spathula*
Pallid sturgeon	*Scaphirhynchus albus*
Plains darter	*Etheostoma pulchellum*
Plains minnow	*Hybognathus placitus*
Plains stoneroller	*Campostoma plumbeum*
Red shiner	*Cyprinella lutrensis*
River shiner	*Notropis blennius*
Sand shiner	*Notropis stramineus*
Shovelnose sturgeon	*Scaphirhynchus platorynchus*
Slender madtom	*Noturus exilis*
Southern redbelly dace	*Chrosomus erythrogaster*
Stonecat	*Noturus flavus*
Suckermouth minnow	*Phenacobius mirabilis*
Topeka shiner	*Notropis topeka*

Amphibians

Plains spadefoot	*Spea bombifrons*

Reptiles

Prairie rattlesnake	*Crotalus viridis*
Texas horned lizard	*Phrynosoma cornutum*
Water moccasin	*Agkistrodon piscivorus*

Birds

American avocet	*Recurvirostra americana*
American golden plover	*Pluvialis dominica*
American goldfinch	*Spinus tristis*
American kestrel	*Falco sparverius*
American redstart	*Setophaga ruticilla*
American robin	*Turdus migratorius*
Baird's sandpiper	*Calidris bairdii*
Bald eagle	*Haliaeetus leucocephalus*
Baltimore oriole	*Icterus galbula*
Black-and-white warbler	*Mniotilta varia*
Blackburnian warbler	*Dendroica fusca*
Black-capped chickadee	*Poecile atricapillus*
Black-headed grosbeak	*Pheucticus melanocephalus*
Black-necked stilt	*Himantopus mexicanus*
Blackpoll warbler	*Dendroica striata*
Black tern	*Chlidonias niger*
Blue jay	*Cyanocitta cristata*
Brown creeper	*Certhia americana*
Brown-headed cowbird	*Molothrus ater*
Buff-breasted sandpiper	*Tryngites subruficollis*
Bullock's oriole	*Icterus bullockii*
Carolina parakeet	*Conuropsis carolinensis*
Carolina wren	*Thyrothorus ludovicianus*
Cassin's sparrow	*Peucaea cassini*
Cedar waxwing	*Bombycilla cedrorum*
Chestnut-collared longspur	*Calcarius ornatus*
Chestnut-sided warbler	*Dendroica pensylvanica*
Chipping sparrow	*Spizella passerina*
Common grackle	*Quiscalus quiscula*
Common raven	*Corvus corax*
Common yellowthroat	*Geothlypis trichas*
Dunlin	*Calidris alpina*
Eastern kingbird	*Tyrannus tyrannus*
Eastern meadowlark	*Sturnella magna*

Eastern phoebe	*Sayornis phoebe*
Eastern whip-poor-will	*Caprimulgus vociferus*
Eastern wood-pewee	*Contopus virens*
Eskimo curlew	*Numenius borealis*
Eurasian skylark	*Aludia arvensis*
Ferruginous hawk	*Buteo regalis*
Franklin's gull	*Leucophaeus pipixcan*
Golden eagle	*Aquila chrysaetos*
Gray catbird	*Dumetella carolinensis*
Greater prairie-chicken	*Tympanuchus cupido*
Greater yellowlegs	*Tringa melancoleuca*
Harris's sparrow	*Zonotrichia querula*
Hermit thrush	*Catharus guttatus*
Horned lark	*Eremophila alpestris*
House wren	*Troglodytes aedon*
Hudsonian godwit	*Limosa haemastica*
Indigo bunting	*Passerina cyanea*
Killdeer	*Charadrius vociferus*
Lapland longspur	*Calcarius lapponicus*
Lark bunting	*Calamospiza melanocorys*
Lazuli bunting	*Passerina ameona*
Least sandpiper	*Calidris minutilla*
Least tern	*Sternula antillarum*
Lesser yellowlegs	*Tringa flavipes*
Long-billed dowitcher	*Limnodromus scolopaceus*
Marbled godwit	*Limosa fedoa*
McCown's longspur	*Rhynochophanes mccownii*
Merlin	*Falco columbarius*
Mountain plover	*Charadrius montanus*
Mourning dove	*Zenaida macroura*
Nashville warbler	*Oreothlypis ruficapilla*
Northern cardinal	*Cardinalis cardinalis*
Northern flicker	*Colaptes auratus*
Northern mockingbird	*Mimus polglattos*
Orange-crowned warbler	*Oreothlypis celata*
Ovenbird	*Seiurus aurocapilla*

Palm warbler	*Dendroica palmarum*
Pectoral sandpiper	*Calidris melanotos*
Peregrine falcon	*Falco peregrinus*
Piping plover	*Charadrius melodus*
Prairie falcon	*Falco mexicanus*
Red-tailed hawk	*Buteo jamaicensis*
Rock wren	*Salpinctes obsoletus*
Rose-breasted grosbeak	*Pheucticus ludovicianus*
Rough-legged hawk	*Buteo lagopus*
Ruby-crowned kinglet	*Regulus calendula*
Sandhill crane	*Grus canadensis*
Say's phoebe	*Sayornus saya*
Scissor-tailed flycatcher	*Tyrannus forficatus*
Semipalmated plover	*Charadrius semipalmatus*
Semipalmated sandpiper	*Calidris pusilla*
Snow goose	*Chen caerulescens*
Sprague's pipit	*Anthus spragueii*
Stilt sandpiper	*Calidris himantopus*
Swainson's hawk	*Buteo swainsoni*
Tennessee warbler	*Oreothlypis peregrina*
Upland sandpiper	*Bartramia longicauda*
Western burrowing owl	*Athene cunicularia*
Western kingbird	*Tyrannus verticalis*
Western meadowlark	*Sturnella neglecta*
Western sandpiper	*Calidris mauri*
Western wood-pewee	*Contopus sordidulus*
White-breasted nuthatch	*Sitta carolinensis*
White-rumped sandpiper	*Calidris fusicollis*
Willet	*Tringa semipalmata*
Wilson's phalarope	*Phalaropus tricolor*
Wilson's snipe	*Gallinago delicata*
Worm-eating warbler	*Helmitheros vermivorum*
Yellow-breasted chat	*Icteria virens*
Yellow-headed blackbird	*Xanthocephalus xanthocephalus*
Yellow-rumped warbler	*Dendroica coronata*
Yellow warbler	*Dendroica petechia*

Mammals

American badger	*Taxidea taxus*
American beaver	*Castor canadensis*
American bison	*Bos bison*
Black-footed ferret	*Mustela nigripes*
Black-tailed prairie dog	*Cynomys ludovicianus*
Columbian mammoth	*Mammuthus columbi*
Coyote	*Canis latrans*
Elk	*Cervus canadensis*
Gray wolf	*Canis lupus*
Mountain lion	*Puma concolor*
Mule deer	*Odocoileus hemionus*
Northern river otter	*Lontra canadensis*
Plains pocket gopher	*Geomys bursarius*
Pronghorn	*Antilocapra americana*
Swift fox	*Vulpes velox*
White-tailed deer	*Odocoileus virginianus*

Selected Bibliography

Prologue

Baughman, Robert. *Kansas in Maps*. Topeka: Kansas State Historical Society, 1961.

Berry, Wendell. *Life Is a Miracle: An Essay against Modern Superstition*. Washington, DC: Counterpoint, 2000.

Heat-Moon, William Least. *PrairyErth: A Deep Map*. Boston: Houghton Mifflin, 1991.

Lewis, Meriwether, William Clark, et al. *The Journals of the Lewis and Clark Expedition*, ed. Gary Moulton. Lincoln, NE: University of Nebraska Press / University of Nebraska-Lincoln Libraries-Electronic Text Center, 2005. https://lewisandclarkjournals.unl.edu/.

Louv, Richard. *Last Child in the Woods: Saving Our Children from Nature-Deficit Disorder*. New York: Algonquin Books of Chapel Hill, 2006.

Shortridge, James. 1980. "Vernacular Regions in Kansas." *American Studies* 21 (Spring 1980): 73–94.

Chapter 1. Headlands

Bettis, Arthur, III, Daniel Muhs, Helen Roberts, and Ann Wintle. "Last Glacial Loess in the Conterminous USA." *Quaternary Science Reviews* 22 (September 2003): 1907–1946.

Buchanan, Rex, ed. *Kansas Geology: An Introduction to Landscapes, Rocks, Minerals, and Fossils*. 2nd ed., rev. Lawrence: University Press of Kansas, 2010.

Cariveau, Allison, and Lacrecia Johnson. *Assessment and Conservation of Playas in Eastern Colorado*. Neotropical Migratory Bird Conservation Act Final Report to the United States Fish and Wildlife Service. Brighton, CO: Rocky Mountain Bird Observatory, 2007.

Carson. Rachel. *The Edge of the Sea*. Boston: Houghton Mifflin, [1955] 1998.

Elias, Maxim. *The Geology of Wallace County, Kansas*. Geological Survey of Kansas Bulletin 18, 1931.

Evans, Catherine. *Playas in Kansas and the High Plains*. Kansas Geological Survey Public Information Circular 30. Lawrence: Kansas Geological Survey, 2010.

Fenneman, Nevin. *Physiography of the Western United States*. New York: McGraw-Hill, 1931.

Frye, John. "The High Plains Surface in Kansas." *Transactions of the Kansas Academy of Science* 49 (June 1946): 71–86.

Frye, John, and O. S. Fent. *The Late Pleistocene Loesses of Central Kansas*. Kansas Geological Survey Bulletin 70, pt. 3. Lawrence: Kansas Geological Survey, 1947.

Greene, Jerome. "The Coming Home Trail." *Kansas History* 38 (Winter 2015-2016): 223–227.

Johnson, Willard. *The High Plains and their Utilization*. Twenty-First Annual Report of the United States Geological Survey, 1899–1900, Pt. 4. Hydrology, 609–732. Washington, DC: Government Printing Office, 1901.

Madole, Richard. "Spatial and Temporal Patterns in Late Quaternary Eolian Deposition, Eastern Colorado, U.S.A." *Quaternary Science Reviews* 14 (1995): 155–177.

Mahler, Harmon, George Engelmann, and Robert Shuster. *Roadside Geology of Nebraska*. Missoula, MT: Mountain, 2003.

Muhs, Daniel. "Age and Paleoclimatic Significance of Holocene Sand Dunes in Northeastern Colorado." *Annals of the Association of American Geographers* 75 (December 1985): 566–582.

———. "Evaluation of Simple Geochemical Indicators of Aeolian Sand Provenance: Late Quaternary Dune Fields of North America Revisited." *Quaternary Science Reviews* 171 (September 2017): 260–296.

Muhs, Daniel, and Vance Holliday. "Evidence of Active Dune Sand on the Great Plains in the 19th Century from Accounts of Early Explorers." *Quaternary Research* 43 (March 1995): 198–208.

Muhs, Daniel, Arthur Bettis III, John Aleinikoff, John McGeehin, Jossh Beann, Gary Skipp, Brian Marshall, Helen Roberts, William Johnson, and Rachel Benton. "Origin and Paleoclimatic Significance of Late Quaternary Loess in Nebraska: Evidence from Stratigraphy, Chronology, Sedimentology, and Geochemistry." *GSA Bulletin* 120 (November/December 2008): 1378–1407.

Muhs, Daniel, John Aleinikoff, Thomas Stafford Jr., Rolf Kihl, Josh Been, Shannon Mahan, and Scott Cowherd. "Late Quaternary Loess in Northeastern Colorado: Part I—Age and Paleoclimatic Significance." *GSA Bulletin* 111 (December 1999): 1861–1875.

Muhs, Daniel, Thomas Stafford, Scott Cowherd, Shannon Mahan, Rolf Kihl, Paul Maat, Charles Bush, and Jennifer Nehring. "Origin of the Late Quaternary Dune Fields of Northeastern Colorado." *Geomorphology* 17 (September 1996): 129–149.

Pearl, Richard. "Pliocene Drainage of East-Central Colorado and Northwestern Kansas." *Mountain Geologist* 8 (1971): 25–30.

Trabert, Sarah, Matthew Hill Jr., and Margaret Beck. "Understanding the Scott County Pueblo (14 SC1) Occupation." In *People in a Sea of Grass: Archaeology's Changing Perspective on Indigenous Plains Communities*, ed. Matthew Hill Jr. and Lauren Ritterbush, 131–143. Salt Lake City: University of Utah Press, 2022.

Velez, Alex. "The Wind Cries Mary": The Effects of Soundscape on the Prairie Madness Phenomenon." *Historical Archaeology* 56 (March 2022): 262–273.

Weaver, John. "Comparison of Vegetation of Kansas-Nebraska Drift-Loess Hills and Loess Plains." *Ecology* 41 (January 1960): 63–88.

Chapter 2. Heartland

Adams, George. "Physiographic Divisions of Kansas." *Transactions of the Kansas Academy of Science* 18 (1903): 109–123.

Blakeslee, Donald. *Holy Ground, Healing Water: Cultural Landscape at Waconda Lake, Kansas*. College Station: Texas A&M University Press, 2010.

Buchanan, Rex, Burke Griggs, and Joshua Svaty. *Petroglyphs of the Smoky Hills*. Lawrence: University Press of Kansas, 2019.

Cather, Willa. "The Enchanted Bluff." *Harper's Monthly Magazine* 118 (April 1909): 774–781.

Copley, Josiah. 1867. *Kansas and the Country Beyond, on the Line of the Union Pacific Railway, Eastern Division, from the Missouri to the Pacific Ocean* . . . Philadelphia, PA: J. B. Lippincott & Co., 1867.

Hattin, Donald. *Stratigraphy and Depositional Environments of Greenhorn Limestone (Upper Cretaceous) of Kansas*. Kansas Geological Survey Bulletin 209. Lawrence: Kansas Geological Survey, 1975.

———. *Stratigraphy and Depositional Environment of Smoky Hill Chalk Member, Niobrara Chalk (Upper Cretaceous) of the Type Area, Western Kansas*. Kansas Geological Survey Bulletin 225. Lawrence: Kansas Geological Survey, 1982.

Kansas Geological Survey website. https://geokansas.ku.edu/.

Lindquist, Emory. *Birger Sandzén: An Illustrated Biography*. Lindsborg, KS: Birger Sandzén Memorial Foundation, 1993.

Muilenberg, Grace, and Ada Swineford. *Land of the Post Rock: Its Origins, History, and People*. Lawrence: University Press of Kansas, 1975.

North, Cori Sherman. *Birger Sandzén: Celebrating the Vision*. Lindsborg, KS: Birger Sandzén Memorial Gallery, 2021.

O'Neill, Brian. *Kansas Rock Art*. Topeka: Kansas State Historical Society, [1981] 1988.

Parks, Douglas, and Waldo Wedel. "Pawnee Geography: Historical and Sacred." *Great Plains Quarterly* 5 (Summer 1985):143–176.

Sandzén, Birger. "The Southwest as a Sketching Ground." *Fine Arts Journal* (August 1915): 333–344, 346–352.

Schoewe, Walter. "The Geographical Center of the United States." *Transactions of the Kansas Academy of Science* 43 (March 1940): 305–306.

———. "Kansas and the Geodetic Datum of North America." *Transactions of the Kansas Academy of Science* 51 (March 1948): 117–124.

———. "The Geography of Kansas: Part II: Physical Geography." *Transactions of the Kansas Academy of Science* 52 (September 1949): 261–333.

Stone, Jeffrey. "Kansas 'Dis-Centers': The Competition to Claim Ownership of the Center of the Nation." *Kansas History* 39 (Spring 2016): 48–63.

Chapter 3. Scarplands

Basinger, James, and David Dilcher. "Ancient Bisexual Flowers." *Science* 224 (May 1984): 511–513.

Bass, Nathan. *Geology of Cowley County, Kansas*. Kansas Geological Survey Bulletin 12. Lawrence: Kansas Geological Survey, 1929.

Beckemeyer, Roy, and Joseph Hall. "The Entomofauna of the Lower Permian Fossil Insect Beds of Kansas and Oklahoma, USA." *African Invertebrates* 48 (April 2007): 23–39.

Charlton, John. "Across the Years on Mount Oread and around the Kaw and Wakarusa River Valleys." *Transactions of the Kansas Academy of Science* 105 (April 2002): 1–17.

Condra, George. *Correlations of the Big Blue Series in Nebraska*. Nebraska Geological Survey Bulletin 6, 2nd ser. Lincoln: Nebraska Geological Survey, 1931.

Dilcher, David, and Peter Crane. "In Pursuit of the First Flower." *Natural History Magazine* 93 (March 1984): 56–61.

Dort, Wakefield, Jr. "Salient Aspects of the Terminal Zone of Continental Glaciation in Kansas." In *Quaternary Environments of Kansas*, ed. William Johnson, 55–66. Kansas Geological Survey Guidebook Series 5. Lawrence: Kansas Geological Survey, 1987.

Everhart, Michael. "Rediscovery of the *Hesperornis regalis* Marsh 1871 Holotype Locality Indicates an Earlier Stratigraphic Occurrence." *Transactions of the Kansas Academy of Science* 114 (2011): 59–68.

———. *Oceans of Kansas: A Natural History of the Western Interior Sea*, 2nd ed. Bloomington: Indiana University Press, 2017.

Frye, John. "The Erosional History of the Flint Hills." *Transactions of the Kansas Academy of Science* 58 (Spring 1955): 79–86.

Harksen, John. "*Pteranodon sternbergi*, a New Pterodactyl from the Niobrara Cretaceous of Kansas." *Proceedings of the South Dakota Academy of Science* 45 (1966): 74–77.

Hart, Stephan, and Archer Hulbert, eds. *The Southwestern Journals of Zebulon Pike, 1806–1807*. Albuquerque: University of New Mexico Press, 2006.

Lewis, Meriwether, William Clark, et al. *The Journals of the Lewis and Clark Expedition*, ed. Gary Moulton. Lincoln, NE: University of Nebraska Press / University of Nebraska-Lincoln Libraries-Electronic Text Center, 2005. https://lewisandclarkjournals.unl.edu/.

Lyle, Shane. *Glaciers in Kansas*. Kansas Geological Survey Public Information Circular 28. Lawrence: Kansas Geological Survey, 2009.

Manchester, Steven, David Dilcher, Walter Judd, Brandon Corder, and James Basinger. "Early Eudicot Flower and Fruit: *Dakotanthus* gen. nov. from the Cretaceous Dakota Formation of Kansas and Nebraska, USA." *Acta Paleobotanica* 58 (2018): 27–40.

Merriam, Daniel. "Southern Extent of Kansan Glaciation (Pleistocene) in Douglas County, Kansas." *Transactions of the Kansas Academy of Science* 106 (April 2003): 17–28.

Rogers, Katherine. *The Sternberg Fossil Hunters: A Dinosaur Dynasty*. Missoula, MT: Mountain, 1991.

Sellards, Elias. "Discovery of Insect Fossils in the Permian of Kansas." *American Journal of Science* 16 (October 1903): 323–324.

Shortridge, James. *Kaw Valley Landscapes: A Traveler's Guide to Northeastern Kansas*. Lawrence: University Press of Kansas, 1988.

Sternberg, Charles. *The Life of a Fossil Hunter*. Bloomington: Indiana University Press, [1909] 1990.

Walker, Myrl. "The Impossible Fossil—Revisited." *Transactions of the Kansas Academy of Science* 109 (Spring 2006): 87–96.

Wright, Ronald. *A Short History of Progress*. Toronto, ON: House of Anansi, 2004.

Chapter 4. Tree Folk

Cather, Willa. *My Ántonia*. Boston: Houghton Mifflin, [1918] 1988.

Grinnell, George Bird. "Cheyenne Stream Names." *American Anthropologist* 8 (January–March 1906): 15–22.

Haddock, Michael, and Craig Freeman. *Trees, Shrubs, and Woody Vines in Kansas*, rev. and exp. ed. Lawrence: University Press of Kansas, 2019.

Hyde, George. *Red Cloud's Folk: A History of the Oglala Sioux Indians*. Norman: University of Oklahoma Press, [1937] 1957.

———. *Life of George Bent Written from His Letters*, ed. Savoie Lottinville. Norman: University of Oklahoma Press, 1968.

James, Edwin, comp. *Account of an Expedition from Pittsburgh to the Rocky Mountains, Performed in the Years 1819 and '20 . . . under the Command of Major Stephen H. Long*. Philadelphia, PA: H. C. Carey and I. Lea, 1823.

Kurz, Don. *Trees of Missouri*. Jefferson City: Conservation Commission of the State of Missouri, 2003.

Nelson, Paul. *The Terrestrial Natural Communities of Missouri*, rev. ed. Jefferson City: Missouri Natural Areas Committee, 2005.

Nuzzo, Victoria. "Extent and Status of Midwest Oak Savanna: Presettlement and 1985." *Natural Areas Journal* 6 (April 1986): 6–36.

Peattie, Donald, and Paul Landacre. *A Natural History of Western Trees*. New York: Bonanza, 1953.

Ware, Eugene. *The Indian War of 1864*. Lincoln: University of Nebraska Press (Bison Book Edition), [1911] 1994.

Wells, Paul. "Scarp Woodlands, Transported Grassland Soils, and Concept of Grassland Climate in the Great Plains Region." *Science* 148 (1965): 246–249.

West, Elliott. *The Contested Plains: Indians, Goldseekers, and the Rush to Colorado*. Lawrence: University Press of Kansas, 1998.

Chapter 5. Setting Sail on the Star-Grass Sea

Albertson, Fredrick. "Ecology of Mixed Prairie in West Central Kansas." *Ecological Monographs* 7 (October 1937): 481–547.

Anderson, Roger. "Evolution and Origin of the Central Grassland of North America: Climate, Fire, and Mammalian Grazers." *Journal of the Torrey Botanical Society* 133 (October–December 2006): 626–647.

Bone, Michael, Dan Johnson, Panayoti Kelaidis, Mike Kintgen, and Larry Vickerman. *Steppes: The Plants and Ecology of the World's Semi-Arid Regions*. Portland, OR: Timber, 2015.

Cather, Willa. *O Pioneers!* Boston: Houghton Mifflin, [1913] 1988.

———. *My Ántonia*. Boston: Houghton Mifflin, [1918] 1988.

Clements, Fredrick. *Plant Indicators: The Relation of Plant Communities to Process and Practice*. Carnegie Institute of Washington Publication 290. Washington, DC: Carnegie Institution for Science, 1920.

Costello, David. *The Prairie World*. Minneapolis: University of Minnesota Press, [1969] 1980.

Davis, W. W. H. *El Gringo: New Mexico and Her People*. Lincoln: University of Nebraska Press (Bison Book Edition), [1857] 1982.

Farrar, Jon. "Sandsage Prairie: A History of the Land." *NEBRASKAland* 71 (September 1993): 22–29.

———. "Sandsage Prairie: The Cinderella Sandhills." *NEBRASKAland* 71 (September 1993): 30–41.

Hanberry, Brice. "Wind-Bounded Grasslands of North America." *Ecological Indicators* 129 (October 2021): 107925.

Heitschmidt, Rodney, G. K. Hulett, and G. W. Tomanek. "Vegetation Map and Community Structure of a West Central Kansas Prairie." *Southwestern Naturalist* 14 (January 1970): 337–350.

Hulett, G. K., C. D. Sloan, and G. W. Tomanek. "The Vegetation of Remnant Grasslands in the Loessial Region of Northwestern Kansas and Southwestern Nebraska." *Southwestern Naturalist* 13 (December 1968): 377–391.

Jackson, Wes. *Becoming Native to this Place*. Washington, DC: Counterpoint, 1996.

Kuchler, August. "A New Vegetation Map of Kansas." *Ecology* 55 (Spring 1974): 586–604.

Larson, Floyd. "The Role of Bison in Maintaining the Short Grass Plains." *Ecology* 21 (April 1940): 113–121.

Lauenroth, William, and Ingrid Burke, eds. *Ecology of the Shortgrass Steppe: A Long-Term Perspective*. New York: Oxford University Press, 2008.

Lauver, Chris, Kelly Kindscher, Don Faber-Langendoen, and Rick Schneider. "A Classification of the Natural Vegetation of Kansas." *Southwestern Naturalist* 44 (December 1999): 421–443.

Locklear, James. "Sandsage Prairie: Floristics, Structure, and Dynamics of a Great Plains Plant Community." *Journal of the Botanical Research Institute of Texas* 13 (July 2019): 253–278.

Magoffin, Susan. *Down the Santa Fe Trail and into Mexico: The Diary of Susan Shelby Magoffin, 1846–1847*, ed. Stella Drumm. Lincoln: University of Nebraska Press, 1982.

Malin, James. *History and Ecology: Studies of the Grassland*, ed. Robert P. Swierenga. Lincoln: University of Nebraska Press, 1984.

Morton, LaDene. *The Waldo Story: The Home of Friendly Merchants*. Charleston, SC: History Press, 2012.

Quinn, James, Daniel Mowrey, Stephen Emanuele, and Ralph Whalley. "The 'Foliage is the Fruit' Hypothesis: *Buchloe dactyloides* (Poaceae) and the Shortgrass Prairie of North America." *American Journal of Botany* 81 (December 1994): 1545–1554.

Ramaley, Francis. "Sand-Hill Vegetation of Northeastern Colorado." *Ecological Monographs* 9 (January 1939): 1–51.

Reichman, O. J. *Konza Prairie: A Tallgrass Natural History*. Lawrence: University Press of Kansas, 1987.

Rolfsmeier, Steven, and Gerry Steinauer. *Terrestrial Ecological Systems and Natural Communities of Nebraska*. Lincoln: Nebraska Natural Heritage Program, Nebraska Game and Parks Commission, 2010.

Schroeder, Walter. *Presettlement Prairie of Missouri*, 2nd ed. Natural History Series no. 2. Jefferson City: Missouri Department of Conservation, 1982.

———. "The Presettlement Prairie in the Kansas City Region (Jackson County, Missouri)." *Missouri Prairie Journal* 6 (December 1985): 3–12.

Shantz, Homer. *Natural Vegetation as an Indicator of the Capabilities of Land for Crop Production in the Great Plains Area*. USDA Bureau of Plant Industry Bulletin 201. Washington, DC: US Department of Agriculture, 1911.

Shortridge, James. *Kansas City and How It Grew, 1822–2011*. Lawrence: University Press of Kansas, 2012.

Tobey, Ronald. *Saving the Prairies: The Life Cycle of the Founding School of American Plant Ecology, 1895–1955*. Berkley: University of California Press, 1981.

Weaver, John. *North American Prairie*. Lincoln, NE: Johnsen, 1954.

Weaver, John, and Fredrick Albertson. *Grasslands of the Great Plains: Their Nature and Use*. Lincoln, NE: Johnsen, 1956.

Wood, Dean. *The Old Santa Fe Trail from the Missouri River*. Kansas City, MO: E. L. Mendenhall, 1955.

Chapter 6. A Kaw Florilegium

Ackerfield, Jennifer. *Flora of Colorado*. Fort Worth: Botanical Research Institute of Texas, 2015.

Bare, Janét. *Wildflowers and Weeds of Kansas*. Lawrence: Regents Press of Kansas, 1979.

Barker, William. "The Flora of the Kansas Flint Hills." *Kansas Science Bulletin* 48 (1969): 525–584.

Barr, Claude. *Jewels of the Plains: Wildflowers of the Great Plains Grasslands and Hills*, rev. ed., ed. James Locklear. Minneapolis: University of Minnesota Press, [1983] 2015.

Berry, Wendell. *A Timbered Choir: The Sabbath Poems 1979–1997*. Washington, DC: Counterpoint, 1998.

Borland, Hal. *High, Wide, and Lonesome*. Tucson: University of Arizona Press, 1956.

Chesterton, G. K. *The Everlasting Man*. San Francisco: Ignatius, [1925] 1993.

Frémont, John Charles. *Report of the Exploring Expedition to the Rocky Mountains* in the Year 1842 and to Oregon and North California in the Years 1843-'44. Washington, DC: Blair & Rives, 1845. Reprint, Washington, DC: Smithsonian Institution, 1988.

———. *Memoirs of My Life: Including in the Narrative Five Journeys of Western Exploration, During the Years 1842, 1843-4, 1845-6-7, 1848-9, 1853-4.* . . . Chicago, IL: Belford, Clarke, 1887.

Geyer, Charles. "Notes on the Vegetation and General Character of the Missouri and Oregon Territories, Made During a Botanical Journey from the State of Missouri, Across the South-Pass of the Rocky Mountains, to the Pacific, During the Years 1843 and 1844." *London Journal of Botany* 4 (1845): 479-492.

Great Plains Flora Association. *Atlas of the Flora of the Great Plains*. Ames: Iowa State University Press, 1977.

Great Plains Flora Association. *Flora of the Great Plains*. Lawrence: University Press of Kansas, 1986.

Haddock, Michael, Craig Freeman, and Janét Bare. *Kansas Wildflowers and Weeds*. Lawrence: University Press of Kansas, 2015.

Kaul, Robert, David Sutherland, and Steve Rolfsmeier. *The Flora of Nebraska*, 2nd ed. Lincoln: School of Natural Resources, University of Nebraska-Lincoln, 2011.

Kindscher, Kelly. *Edible Wild Plants of the Prairie: An Ethnobotanical Guide*. University Press of Kansas, 1987.

Kindscher, Kelly, and Paul Wells. "Prairie Plant Guilds: A Multivariate Analysis of Prairie Species Based on Ecological and Morphological Traits." *Vegetatio* 117 (March 1995): 29-50.

Locklear, James. *Phlox: A Natural History and Gardener's Guide*. Portland, OR: Timber, 2011.

Majetic, Cassie, Shelly Wiggam, and Carolyn Ferguson. "Timing Is Everything: Temporal Variation in Floral Scent, and Its Connections to Pollinator Behavior and Female Reproductive Success in *Phlox divaricata*." *American Midland Naturalist* 173 (April 2015): 191-207.

Mattes, Merrill. *The Great Platte River Road*, 2nd ed. Lincoln: University of Nebraska Press, [1969] 1987.

Ramaley, Francis. "Sand-Hill Vegetation of Northeastern Colorado." *Ecological Monographs* 9 (January 1939): 1-51.

Stevens, William. *Kansas Wildflowers*, 2nd ed. Lawrence: University of Kansas Press, 1948.

Weaver, John. *The Ecological Relations of Roots*. Carnegie Institute of Washington Publication 286. Washington, DC: Carnegie Institute, 1919.

Weaver, John, and T. J. Fitzpatrick. "The Prairie." *Ecological Monographs* 4 (April 1934): 109-295.

Weber, William A., and Ronald C. Wittmann. *Colorado Flora, Eastern Slope: A Field Guide to the Vascular Plants*, 4th ed. Boulder: University Press of Colorado, 2012.

Welsh, Stanley. *John Charles Frémont: Botanical Explorer*. St. Louis: Missouri Botanical Garden, 1998.

Wiggam, Shelly, and Carolyn Ferguson. "Pollinator Importance and Temporal Variation in a Population of *Phlox divaricata* L. (Polemoniaceae)." *American Midland Naturalist* 154 (July 2005): 42–54.

Wislizenus, Friedrich. "Memoir of a Tour to Northern Mexico, Connected with Col. Doniphan's Expedition in 1846 and 1847." United States 30th Cong., 1st Sess. misc. no. 26. Washington, DC: Tippin & Streeper, 1848.

Chapter 7. The Days of Manure River

Albertson, Fredrick. "Ecology of Mixed Prairie in West Central Kansas." *Ecological Monographs* 7 (October 1937): 481–547.

Blakeslee, Donald. *Holy Ground, Healing Water: Cultural Landscape at Waconda Lake, Kansas*. College Station: Texas A&M University Press, 2010.

Chalfant, William. *Cheyennes and Horse Soldiers: The 1857 Expedition and the Battle of Solomon's Fork*. Norman: University of Oklahoma Press, 1989.

———. *Cheyennes at Dark Water Creek: The Last Fight of the Red River War*. Norman: University of Oklahoma Press, 1997.

Dodge, Richard. *The Hunting Grounds of the Great West: A Description of the Plains, Game, and Indians of the Great North American Desert*. London: Chatto & Windus, 1877.

Fitch, Julian. "Lieut. Fitch's Report on the Smoky Hill Route." In Eugene F. Ware, *The Indian War of 1864*, 419–425. Lincoln: University of Nebraska Press (Bison Book Edition), [1911] 1994.

Flores, Dan. "Bison Ecology and Bison Diplomacy: The Southern Plains from 1800 to 1850." *Journal of American History* 78 (September 1991): 465–485.

———. *American Serengeti: The Last Big Animals of the Great Plains*. Lawrence: University Press of Kansas, 2016.

Frémont, John Charles. *Report of the Exploring Expedition to the Rocky Mountains in the Year 1842 and to Oregon and North California in the Years 1843-'44*. Washington, DC: Blair & Rives, 1845. Reprint, Washington, DC: Smithsonian Institution, 1988.

Frye, John. "The High Plains Surface in Kansas." *Transactions of the Kansas Academy of Science* 49 (June 1946): 71–86.

Grinnell, George Bird. *The Fighting Cheyennes*. Norman: University of Oklahoma Press, [1915] 1956.
Hart, Stephan, and Archer Hulbert, eds. *The Southwestern Journals of Zebulon Pike, 1806–1807*. Albuquerque: University of New Mexico Press, 2006.
Hyde, George. *The Pawnee Indians*. Norman: University of Oklahoma Press, 1951.
Larson, Floyd. "The Role of Bison in Maintaining the Short Grass Plains." *Ecology* 21 (April 1940): 113–121.
Mattes, Merrill, ed. "Capt. L. C. Easton's Report: Fort Laramie to Fort Leavenworth via Republican River in 1849." *Kansas Historical Quarterly* 20 (May 1953): 392–417.
Norall, Frank. *Bourgmont: Explorer of the Missouri, 1698–1725*. Lincoln: University of Nebraska Press, 1988.
Roper, Donna. "John Dunbar's Journal of the 1834–5 Chawi Winter Hunt and Its Implications for Pawnee Archaeology." *Plains Anthropologist* 36 (August 1991): 193–214.
———. "Documentary Evidence for Changes in Protohistoric and Early Historic Pawnee Hunting Practices." *Plains Anthropologist* 37 (August 1992): 353–366.
Sandoz, Mari. *The Buffalo Hunters: The Story of the Hide Men*, 2nd ed. Lincoln: University of Nebraska. Press (Bison Book Edition) [1954] 2008.
———. *Love Song to the Plains*. New York: Harper & Brothers, 1961.
Scott, Douglas, Peter Bleed, and Stephen Damm. *Custer, Cody, and Grand Duke Alexis: Historical Archaeology of the Royal Buffalo Hunt*. Norman: University of Oklahoma Press, 2013.
West, Elliott. *The Contested Plains: Indians, Goldseekers, and the Rush to Colorado*. Lawrence: University Press of Kansas, 1998.
Woodhouse, Connie, Jeffrey Lukas, and Peter Brown. "Drought in the Western Great Plains, 1845–56: Impacts and Implications." *Bulletin of the American Meteorological Society* 83 (October 2002): 1485–1493.

Chapter 8. O Elkader!

Armstrong, David, James Fitzgerald, and Carron Meaney. *Mammals of Colorado*, 2nd ed. Denver: Denver Museum of Nature & Science, 2011.
Baker, A. B. "Mammals of Western Kansas." *Transactions of the Kansas Academy of Science* 11 (1887–1888): 56–58.
Bryan, Rebecca, and Michael Wunder. "Western Burrowing Owls (*Athene cunicularia hypugaea*) Eavesdrop on Alarm Calls of Black-Tailed Prairie Dogs (*Cynomys ludovicianus*)." *Ethology* 120 (February 2014): 180–188.

Chipault, Jennifer, and James Detling. "Bison Selection of Prairie Dog Colonies on Shortgrass Steppe." *Western North American Naturalist* 73 (July 2013): 168–176.

Choate, Jerry, Edward Boggess, and Robert Henderson. "History and Status of the Black-Footed Ferret in Kansas." *Transactions of the Kansas Academy of Science* 85 (1982): 121–132.

Davis, W. W. H. *El Gringo: New Mexico and Her People*. Lincoln: University of Nebraska Press (Bison Book Edition), [1857] 1982.

Flores, Dan. *American Serengeti: The Last Big Animals of the Great Plains*. Lawrence: University Press of Kansas, 2016.

Frazier, George. *The Last Wild Places of Kansas: Journeys into Hidden Landscapes*. Lawrence: University Press of Kansas, 2016.

James, Edwin, comp. *Account of an Expedition from Pittsburgh to the Rocky Mountains, Performed in the Years 1819 and '20 . . . under the Command of Major Stephen H. Long*. Philadelphia, PA: H.C. Carey and I. Lea, 1823.

Johnsgard, Paul. *Prairie Dog Empire: A Saga of the Shortgrass Prairie*: Lincoln: University of Nebraska Press, 2014.

Klataske, Ron. "Conservation of Prairie Dogs and Reintroduction of Black-footed Ferrets Requires Courage." *Prairie Wings* (Fall/Winter 2011): 14–18.

Kotliara, Natasha, Bruce Baker, April Whicker, and Glenn Plumb. "A Critical Review of Assumptions about the Prairie Dog as a Keystone Species." *Environmental Management* 24 (October 1999): 177–192.

Kretzer, Justin, and Jack Cully Jr. "Prairie Dog Effects on Harvester Ant Species Diversity and Density." *Journal of Range Management* 54 (January 2001): 11–14.

———. "Effects of Black-tailed Prairie Dogs on Reptiles and Amphibians in Kansas Shortgrass Prairie." *Southwestern Naturalist* 46 (June 2001): 171–177.

Mead, James. "Notes on Two Kansas Mammals." *Bulletin of the Washburn College Laboratory of Natural History* 1 (1885): 91–92.

———. "Some Natural History Notes of 1859." *Transactions of the Annual Meeting of the Kansas Academy of Sciences* 16 (1897–1898): 280–281.

Paine, Robert. "A Note on Trophic Complexity and Community Stability." *American Naturalist* 103 (January-February 1969): 91–93.

Parker, Ryan, Courtney Duchardt, Angela Dwyer, Cristi Painter, Allison Pierce, Tyler Michels, and Michael Wunder. "Trophic Ecology Warrants Multispecies Management in a Grassland Setting: Proposed Species Interactions on Black-tailed Prairie Dog Colonies." *Rangelands* 41 (June 2019): 135–144.

Roth, Stanley, Jr. and John Marzluff. "Nest Placement and Productivity of Ferruginous Hawks in Western Kansas." *Transactions of the Kansas Academy of Science* 92 (1989): 132–148.

Rydjord, John. *Kansas Place-Names*. Norman: University of Oklahoma Press, 1972.

Scheffer, Theodore. "The Prairie-Dog Situation in Kansas." *Transactions of the Kansas Academy of Science* 23/24 (January 1911): 115–118.

Sexson, Mark, and Jerry Choate. "Historical Biogeography of the Pronghorn in Kansas." *Transactions of the Kansas Academy of Science* 84 (1981): 128–133.

Sexson, Mark, Jerry Choate, and Robert Nicholson. "Diet of Pronghorn in Western Kansas." *Journal of Range Management* 34 (November 1981): 489–493.

Smith, Gregory, and Mark Lomolino. "Black-tailed Prairie Dogs and the Structure of Avian Communities on the Shortgrass Plains." *Oecologia* 138 (April 2004): 592–602.

Uhey, Derek, and Richard Hofstretter. "From Pests to Keystone Species: Ecosystem Influences and Human Perceptions of Harvester Ants (*Pogonomyrmex*, *Veromessor*, and *Messor* spp.)." *Annals of the Entomological Society of America* 115 (March 2022): 127–140.

Vanderhoof, Jennifer, Robert Robel, and Kenneth Kemp. "Numbers and Extent of Black-Tailed Prairie Dog Towns in Kansas." *Transactions of the Kansas Academy of Science* 97 (April 1994): 36–43.

Wishart, David. *The Last Days of the Rainbelt*. Lincoln: University of Nebraska Press, 2013.

Chapter 9. Bird Sketches

Andrews, Robert, and Robert Righter. *Colorado Birds: A Reference to Their Distribution and Habitat*. Denver: Denver Museum of Natural History, 1992.

Audubon, John James, author, artist. *The Birds of America: From Drawings Made in the United States and Their Territories*. New York: J. B. Chevalier, 1840–1844.

Bailey, Alfred, and Robert Niedrach. *Birds of Colorado*, 2 vol. Denver: Denver Museum of Natural History, 1965.

Baumgartner, Frederick, and Marguerite Baumgartner. *Oklahoma Bird Life*. Norman: University of Oklahoma Press, 1992.

Beauvais, Gary, James Enderson, and Anthony Magro. 1992. "Home Range, Habitat Use and Behavior of Prairie Falcons Wintering in East-Central Colorado." *Journal of Raptor Research* 26 (March 1992): 13–18.

Brooking, Albert. "The Vanishing Bird Life of Nebraska." *Nebraska Bird Review* 10 (1942): 43–47.

Cather, Willa. *My Ántonia*. Boston: Houghton Mifflin, [1918] 1988.

Dunn, John, and Kimball Garrett. *A Field Guide to the Warblers of North America*. Boston, MA: Houghton Mifflin, 1997.

Gardner, W. H. and N. H. MacKenzie. *The Poems of Gerard Manley Hopkins*, 4th ed. Oxford: Oxford University Press, 1970.

Gilfillan, Merrill. *Chokecherry Places: Essays from the High Plains*. Boulder, CO: Johnson, 1998.

———. *Rivers & Birds*. Boulder, CO: Johnson, 2003.

———. *The Warbler Road*. Chicago, IL: Flood, 2010.

Goss, Nathaniel. *The History of the Birds of Kansas: Illustrating 529 Birds*. Topeka, KS: George W. Crane, 1891.

Hill, Asa. "Mr. A. T. Hill's Own Story." *Nebraska History Magazine* 10 (1927): 162–167.

Jacobs, Brad. *Birds in Missouri*. Jefferson City: Missouri Department of Conservation, 2001.

Johnsgard, Paul. *Hawks, Eagles, & Falcons of North America: Biology and Natural History*. Washington, DC: Smithsonian Institution, 1990.

———. *Prairie Birds: Fragile Splendor in the Great Plains*. Lawrence: University Press of Kansas, 2001.

Nuttall, Thomas. *A Manual of the Ornithology of the United States and Canada: The Land Birds*, 2nd ed., with additions. Boston: Hilliard, Gray, 1840.

Pickwell, Gayle. "The Prairie Horned Lark." *Transactions of the Academy of Science of St. Louis* 27 (August 1931): 1–153.

Rising, James. *A Guide to the Identification and Natural History of the Sparrows of the United States and Canada*. San Diego, CA: Academic, 1966.

Sherman, Althea. *Birds of an Iowa Dooryard*, ed. Fred Pierce. Iowa City: University of Iowa Press, [1952] 1996.

Thompson, Max, and Charles Ely. *Birds in Kansas*, 2 vol. Lawrence: University Press of Kansas, 1989.

Thompson, Max, Charles Ely, Bob Gress, Chuck Otte, Sebastian Patti, David Seibel, and Eugene Young. *Birds of Kansas*. Lawrence: University Press of Kansas, 2011.

Townsend, John Kirk. *Narrative of a Journey across the Rocky Mountains to the Columbia River*. Corvallis: Oregon State University Press, [1839] 1999.

Zimmerman, John, and Sabastian Patti. *A Guide to Bird Finding in Kansas and Western Missouri*. Lawrence: University Press of Kansas, 1988.

Chapter 10. Surprised by Shorbs

Gillespie, Caitlyn, and Joseph Fontaine. "Shorebird Stopover Habitat Decisions in a Changing Landscape." *Journal of Wildlife Management* 81 (August 2017): 1051–1062.

Johnsgard, Paul. *Plovers, Sandpipers, and Snipes of the World*. Lincoln: University of Nebraska Press, 1981.

———. *Prairie Birds: Fragile Splendor in the Great Plains*. Lawrence: University Press of Kansas, 2001.
Jorgensen, Joel. *An Overview of Shorebird Migration in the Eastern Rainwater Basin, Nebraska*. Nebraska Ornithologists' Union Occasional Paper no. 8. Wakefield, Nebraska Ornithologists' Union, 2004.
———. *Update to an Overview of Shorebird Migration in the Eastern Rainwater Basin, Nebraska*. Published by Author, 2008.
———. *Birds of the Rainwater Basin, Nebraska*. Nebraska Game and Parks Commission Staff Research Publications 55. Lincoln: Nebraska Game and Parks Commission, 2012.
Jorgensen, Joel, John McCarty, and LaReesa Wolfenbarger. "Buff-Breasted Sandpiper Density and Numbers during Migratory Stopover in the Rainwater Basin, Nebraska." *Condor* 110 (February 2008): 63–69.
Lingle, Gary. *Birding Crane River: Nebraska's Platte*. Grand Island, NE: Harrier, 1994.
McCarty, John, Joel Jorgensen, and LaReesa Wolfenbarger. "Behavior of Buff-Breasted Sandpipers (*Tryngites subruficollis*) during Migratory Stopover in Agricultural Fields." *PLoS ONE* 4 (November 2009): 1–5.
McCarty, John, Joel Jorgensen, Justin Michaud, and LaReesa Wolfenbarger. "Buff-Breasted Sandpiper Stopover Duration in the Rainwater Basin, Nebraska, in Relation to the Temporal and Spatial Migration Patterns in the Great Plains of North America." *Wader Study* 122 (December 2015): 243–254.
Morris, Lee. "Buff-Breasted Sandpipers." *Nebraska Bird Review* 46 (December 1978): 77–79.
———. "Notes on Bird Sightings in Nebraska." *Nebraska Bird Review* 63 (June 1995): 60.
Rainwater Basin Joint Venture. https://www.rwbjv.org/.
Rice, Nathan, Kristof Zyskowski, and William Busby. 2001. "Charadiiform Bird Surveys in North Central Kansas." *Kansas Ornithological Society* 52 (September 2001): 29–36.
Senner, Nathan, Wesley Hochachka, James Fox, and Vsevolod Afanasyev. "An Exception to the Rule: Carry-Over Effects Do Not Accumulate in a Long-Distance Migratory Bird." *PLoS ONE* 9 (February 2014): 1–11.
Sharpe, Roger, Ross Silcock, and Joel Jorgensen. *Birds of Nebraska: Their Distribution and Temporal Occurrence*. Lincoln: University of Nebraska Press, 2001.
Webb, Elisabeth, Loren Smith, Mark Vrtiska, Theodore LaGrange. "Community Structure of Wetland Birds during Spring Migration through the Rainwater Basin." *Journal of Wildlife Management* 74 (May 2010): 765–777.
Zimmerman, John. *Cheyenne Bottoms: Wetland in Jeopardy*. Lawrence: University Press of Kansas, 1990.

Chapter 11. The Waters of Mother Kaw

Barry, Louise. "Kansas before 1854: A Revised Annals, Part Nine, 1836–1837." *Kansas Historical Quarterly* 29 (Spring 1963): 41–81.

Brady, Lawrence, comp., David Grisafe, James McCauley, Gregory Ohlmacher, Hernan Qinodoz, and Kenneth Nelson. *The Kansas River Corridor: Its Geologic Setting, Land Use, Economic Geology, and Hydrology*. Kansas Geological Survey Open File Report 98-2. Lawrence: Kansas Geological Survey, 1998.

Breukelman, John. "The Fishes of Northwestern Kansas." *Transactions of the Kansas Academy of Science* 43 (March 1940): 367–375.

Condra, George. *Geology and Groundwater Resources of the Republican River Valley and Adjacent Areas*. US Geological Survey Water Supply Paper 216. Reston, VA: US Geological Survey, 1907.

Costello, David. *The Prairie World*. Minneapolis: University of Minnesota Press, [1969] 1980.

Costigan, Katie, Melinda Daniels, and Walter Dodds. "Fundamental Spatial and Temporal Disconnections in the Hydrology of an Intermittent Prairie Headwater Network." *Journal of Hydrology* 522 (March 2015): 305–316.

Dodds, Walter, Keith Gido, Matt Whiles, Ken Fritz, and William Matthews. "Life on the Edge: The Ecology of Great Plains Prairie Streams." *BioScience* 54 (March 2004): 205–216.

Eberle, Mark. "Type Locality and Conservation Status of the Northern Plains Killifish (*Fundulus kansae*: Fundulidae) in Kansas." *Transactions of the Kansas Academy of Science* 112 (Spring 2009): 87–97.

Eberle, Mark, Guy Ernsting, Bill Stark, and Joseph Tomelleri. "Recent Surveys of Fishes from Western Kansas." *Transactions of the Kansas Academy of Science* 92 (1989): 24–32.

Falke, Jeffrey, Kurt Fausch, Kevin Bestgen, and Larissa Bailey. "Spawning Phenology and Habitat Use in a Great Plains, USA, Stream Fish Assemblage: An Occupancy Estimation Approach." *Canadian Journal of Fisheries and Aquatic Sciences* 67 (November 2010): 1942–1956.

Falke, Jeffrey, Kurt Fausch, Robin Magelky, Angela Aldred, Deanna Durnford, Linda K. Riley, and Ramchand Oad. "The Role of Groundwater Pumping and Drought in Shaping Ecological Futures for Stream Fishes in a Dryland River Basin of the Western Great Plains." *Ecohydrology* 4 (September 2011): 682–697.

Falke, Jeffrey, Larissa Bailey, Kurt Fausch, and Kevin Bestgen. "Colonization and Extinction in Dynamic Habitats: An Occupancy Approach for a Great Plains Stream Fish Assemblage." *Ecology* 93 (April 2012): 858–867.

Flora, Snowdon. "The Great Flood of 1844 along the Kansas and Marais des Cygnes Rivers." *Kansas Historical Quarterly* 20 (May 1952): 73–81.

Gido, Keith, and Walter Dodds. "Retrospective Analysis of Fish Community Changes during a Half-Century of Landuse and Streamflow Changes." *Journal of the North American Benthological Society* 29 (September 2010): 970–987.

Haslouer, Stephen, Mark Eberle, David Edds, Keith Gido, Chris Mammoliti, James Triplett, Joseph Collins, Donald Distler, Donald Huggins, and William Stark. "Current Status of Native Fish Species in Kansas." *Transactions of the Kansas Academy of Science* 108 (2005): 32–46.

Hoagstrom, Christopher, James Brooks, and Stephen Davenport. "A Large-Scale Conservation Perspective Considering Endemic Fishes of the North American Plains." *Biological Conservation* 144 (January 2011): 21–34.

Joseph Tomelleri. https://www.americanfishes.com/en/.

Kansas Fishes Committee. *Kansas Fishes*. Lawrence: University Press of Kansas, 2014.

Martin, Erika, James Whitney, Keith Gido, and Kristen Hase. "Habitat Associations of Stream Fishes in Protected Tallgrass Prairie Streams." *American Midland Naturalist* 170 (July 2013): 39–51.

Minckley, W. L. "A Fish Survey of the Pillsbury Crossing Area, Deep Creek, Riley County, Kansas." *Transactions of the Kansas Academy of Science* 59 (Autumn 1956): 351–357.

Minckley, W. L., and Frank Cross. "Distribution, Habitat, and Abundance of the Topeka Shiner *Notropis topeka* (Gilbert) in Kansas. *American Midland Naturalist* 61 (January 1959): 210–217.

Opie, John, Char Miller, and Kenna Lang Archer. *Ogallala: Water for a Dry Land*, 3rd ed. Lincoln: University of Nebraska Press, 2018.

Perkins, Joshuah, Keith Gido, Jeffrey Falke, Kurt Fausch, Harry Crockett, Eric Johnson, and John Sanderson. "Groundwater Declines Are Linked to Changes in Great Plains Stream Fish Assemblages." *Proceedings of the National Academy of Sciences* 114 (July 2017): 7373–7378.

Prior, Adam, Ramchand Oad, and Kristoph-Deitrich Kinzli. "Agricultural Water Conservation in the High Plains Aquifer and Arikaree River Basin." *Journal of Water Resource and Protection* 5 (January 2013): 747–759.

Reichman, O. J. *Konza Prairie: A Tallgrass Natural History*. Lawrence: University Press of Kansas, 1987.

Scheurer, Julie, Kurt Fausch, and Kevin Bestgen. "Multiscale Processes Regulate Brassy Minnow Persistence in a Great Plains Stream." *Transactions of the American Fisheries Society* 132 (September 2003): 840–855.

Schoewe, Walter. "The Geography of Kansas: Part III: Hydrogeography." *Transactions of the Kansas Academy of Science* 54 (September 1951): 263–329.

Stark, William, Jason Luginbill, and Mark Eberle. "Natural History of a Relict

Population of Topeka Shiner (*Notropis topeka*) in Northwestern Kansas." *Transactions of the Kansas Academy of Science* 105 (October 2002): 143–152.

Stubbs, Stacey. "'Gentle River Goes Mad': The Republican River Flood of 1935 and its New Deal Legacy." *Nebraska History* 97 (Spring 2016): 2–15.

Thompson, Craig. *Along the Kaw: A Journey Down the Kansas River*. Published by Author, 2012.

Tomelleri, Joe, Mark Eberle, and Guy Ernsting. *Big Creek and Its Fishes*. Fort Hays Studies Series No. 75. Fort Hays, KS: Fort Hays State University, 1986.

US Environmental Protection Agency, Office of Water Quality, Region VII. *Everyone Can't Live Upstream: A Contemporary History of Water Quality Problems on the Missouri River, Sioux City, Iowa to Herman, Missouri*. Kansas City, MO: Environmental Protection Agency, 1971.

Chapter 12. Beautiful Contrivances

Darwin, Charles. *On the Origin of Species by Means of Natural Selection, or the Preservation of Favoured Races in the Struggle for Life*, with an Introduction and Notes by George Levine. New York: Barnes & Noble Classics, [1859], 2004.

———. *On the Various Contrivances by which British and Foreign Orchids are Fertilised by Insects, and on the Good Effects of Intercrossing*. London: John Murray, 1862.

Gilmore, Melvin. *Prairie Smoke*. St. Paul: Minnesota Historical Society, [1929] 1987.

Jadeja, Shivani, and Brigitte Tenhumberg. "Phytophagous Insect Oviposition Shifts in Response to Probability of Flower Abortion Owing to the Presence of Basal Fruits." *Ecology and Evolution* 7 (September 2017): 8770–8779.

Kopper, Brian, David Margolies, and Ralph Charlton. "Life History Notes on Regal Fritillary, *Speyeria idalia* (Drury) (Lepidoptera: Nymphalidae), in Kansas Tallgrass Prairie." *Journal of the Kansas Entomological Society* 74 (July 2001): 172–177.

Kopper, Brian, Shengqiang Shu, Ralph Charlton, and Sonny Ramaswamy. "Evidence for Reproductive Diapause in the Fritillary *Speyeria idalia* (Lepidoptera: Nymphalidae)." *Annals of the Entomological Society in America* 94 (May 2001): 427–432.

Leopold, Aldo. *A Sand County Almanac, and Sketches Here and There*. New York: Oxford University Press, [1949] 1968.

Macior, Walter. "The Pollination Ecology of *Dicentra cucullaria*." *American Journal of Botany* 57 (January 1970): 6–11.

McCullough, Kelsey, Gene Albanese, and David Haukos. "Novel Observations of

Larval Fire Survival, Feeding Behavior, and Host Plant Use in the Regal Fritillary, *Speyeria idalia* (Drury) (Nymphalidae)." *Journal of the Lepidopterists' Society* 71 (September 2017): 146–152.

McGinley, Ronald. "Studies of Halictinae (Apoidae: Halictidae), II: Revision of *Sphecodogastra* Ashmead, Floral Specialists of Onagraceae." *Smithsonian Contributions to Zoology* 610 (2003).

Nonnenmacher, Hermann. "The Comparative Floral Ecology of Seven Species of Onagraceae in Native Tallgrass Prairie." PhD diss., Saint Louis University, 1999.

Powell, Alexis, William Busby, and Kelly Kindscher. "Status of Regal Fritillary (*Speyeria idalia*) and Effects of Fire Management on its Abundance in Northeastern Kansas, USA." *Journal of Insect Conservation* 11 (September 2007): 299–308.

Smyth, B. B. "Periodicity in Plants." *Transactions of the Annual Meeting of the Kansas Academy of Science* 12 (1989–1890): 75–81.

Stevens, Orin. "Notes on the Species of *Halictus* Visiting Evening Flowers (Hym.). *Entomological News* 31 (February 1920): 35–44.

Stevens, William. *Kansas Wildflowers*, 2nd ed. Lawrence: University of Kansas Press, 1948.

Svensson, Glenn, Olle Pellmyr, and Robert Raguso. "Pollinator Attraction to Volatiles from Virgin and Pollinated Host Flowers in Yucca/Moth Obligate Mutualism." *Oikos* 120 (October 2011): 1577–1583.

Turnbull, Christine, and David Culver. "The Timing of Seed Dispersal in *Viola nuttallii*: Attraction of Dispersers and Avoidance of Predators." *Oecologia* 59 (1983): 360–365.

Turnbull, Christine, Andrew Beattie, and Frances Hanzawa. "Seed Disperal by Ants in the Rocky Mountains." *Southwestern Naturalist* 28 (August 1983): 289–293.

Chapter 13. The Hard Places

Benedict, Lauryn, and Nathanial Warning. "Rock Wrens Preferentially Use Song Types that Improve Long Distance Signal Transmission during Natural Singing Bouts." *Journal of Avian Biology* 48 (June 2017): 1254–1262.

Borland, Hal. *High, Wide, and Lonesome*. Tucson: University of Arizona Press, 1956.

Collins, Joseph, Suzanne Collins, and Travis Taggart. *Amphibians, Reptiles, and Turtles in Kansas*, Eagle Mountain, UT: Eagle Mountain Publishing, 2010.

Dillard, Annie. *Pilgrim at Tinker Creek*. New York: Harper & Row, 1974.

Eiseley, Loren. *The Immense Journey*. New York: Time Incorporated, [1946] 1962.

———. *All the Strange Hours: The Excavation of a Life*. New York: Charles Scribner's Sons, 1975.

Hammerson, Geoffrey, *Amphibians and Reptiles in Colorado*, 2nd ed. Louisville, CO: University Press of Colorado & Colorado Division of Wildlife, 1999.

Janovy, John, Jr., *Keith County Journal*. Lincoln: University of Nebraska Press (Bison Book Edition), [1978] 1996.

Merola, Michele. "Observations of the Nesting and Breeding Behavior of the Rock Wren." *Condor* 97 (May 1995): 585–587.

O'Neill, Brian. "Community Disassembly in Ephemeral Ecosystems." *Ecology* 97 (December 2016): 3285–3292.

O'Neill, Brian, Christopher Rogers, and James Thorp. "Flexibility of Ephemeral Wetland Crustaceans: Environmental Constraints and Anthropogenic Impacts." *Wetlands Ecological and Management* 24 (August 2015): 279–291.

Pfennig, David. "The Adaptive Significance of an Environmentally-Cued Developmental Switch in an Anuran Tadpole." *Oecologia* 85 (November 1990): 101–107.

———. "Polyphenism in Spadefoot Toad Tadpoles as Locally Adjusted Evolutionarily Stable Strategy." *Evolution* 46 (October 1992): 1408–1420.

Smyth, Michael, and George Bartholomew. "The Water Economy of the Black-Throated Sparrow and the Rock Wren." *Condor* 68 (September 1966): 447–458.

Warning, Nathanial, and Lauryn Benedict. "Paving the Way: Multifunctional Nest Architecture of the Rock Wren." *Auk* 132 (January 2015): 288–299.

———. "Facultative Nest Modification by Rock Wrens (*Salpinctes obsoletus*)." *Avian Biology Research* (March 2016): 58–65.

Chapter 14. Paradise Undone

Behrendt, Stephen. "One Man's Dust Bowl: Recounting 1936 with Don Hartwell of Inavale, Nebraska." *Great Plains Quarterly* 35 (Summer 2015): 229–247.

Berry, Wendell. "A Native Hill." In *The Art of the Commonplace: The Agrarian Essays of Wendell Berry*, ed. Norman Wirzba, 3–31. Washington, DC: Counterpoint, 2002.

Blackmore, William, Introduction to *The Hunting Grounds of the Great West: A Description of the Plains, Game, and Indians of the Great North American Desert*, by Richard Dodge, xv–lvii. London: Chatto & Windus, 1877.

Cook, Benjamin, Ron Miller, and Richard Seager. "Amplification of the North American 'Dust Bowl' Drought through Human-Induced Land Degradation." *Proceedings of the National Academy of Sciences* 106 (March 2009): 4997–5001.

Davis, Kenneth. *River on the Rampage*. Garden City, NY: Doubleday, 1953.

Dodge, Richard. *The Hunting Grounds of the Great West: A Description of the Plains, Game, and Indians of the Great North American Desert*. London: Chatto & Windus, 1877.

Eagan, Timothy. *The Worst Hard Time: The Untold Story of Those Who Survived the Great American Dust Bowl*. Boston, MA: Houghton Mifflin, 2006.

Gregg, Josiah. *Commerce of the Prairies*, ed. Max L. Moorehead. Norman: University of Oklahoma Press, [1844] 1954.

Herron, John. "Making Meat: Race, Labor, and the Kansas City Stockyards." In *Wide-Open Town: Kansas City in the Pendergast Era*, ed. Diane Mutte Burke, Jason Roe, and John Herron, 119–138. Lawrence: University Press of Kansas, 2018.

Hewes, Leslie. "Suitcase Farming in the Central Great Plains." *Agricultural History* 51 (January 1977): 23–27.

Hornaday, William. "The Extermination of the American Bison, with a Sketch of its Discovery and Life History." In *Report of the United States National Museum for the year ending June 30, 1887 (Pt. 2 of the Annual Report of the Board of Regents of the Smithsonian Institution for the year ending June 30, 1887)*, 367–548, 21 plates, 1 map. Washington, DC: Smithsonian, 1889.

Lee, Jeffrey, and Thomas Gill. "Multiple Causes of Wind Erosion in the Dust Bowl." *Aeolian Research* 19 (2015): 15–36.

Mallea, Amahia K. *A River in the City of Fountains: An Environmental History of Kansas City and the Missouri River*. Lawrence: University Press of Kansas, 2018.

Mead, James. "Some Natural History Notes of 1859." *Transactions of the Annual Meeting of the Kansas Academy of Science* 16 (December 1898): 280–281.

———. "The Saline River County in 1859." *Kansas State Historical Society Collections* 9 (1906): 8–19. Quote from p. 15.

———. *Hunting and Trading on the Great Plains 1859–1875*, ed. Schuyler Jones. Wichita, KS: Rowfant, [1986] 2008.

Peck, Robert Morris. *The Wolf Hunters: A Story of the Buffalo Plains*, ed. George Bird Grinnell. New York: Charles Scribner's Sons, 1914.

Priddy, Bob. *Only the Rivers and Peaceful: Thomas Hart Benton's Missouri Mural*. Independence, MO: Independence Press, 1989.

Sandoz, Mari. *The Buffalo Hunters: The Story of the Hide Men*, 2nd ed. Lincoln: University of Nebraska. Press (Bison Book Edition) [1954] 2008.

Serda, Daniel. "A Blow to the Spirit: The Kaw River Flood of 1951 in Perspective." Paper presented at Midcontinent Perspectives lecture series, Midwest Research Institute, Kansas City, MO, October 28, 1993.

Shortridge, James. *Kansas City and How It Grew, 1822–2011*. Lawrence: University Press of Kansas, 2012.

Streeter, Floyd. *The Kaw: The Heart of a Nation*. New York: Farrar & Rinehart, 1941.
Swenk, Myron. "The Eskimo Curlew and Its Disappearance." *Proceedings of the Nebraska Ornithologist's Union* 6 (February 1915): 25–44.
Worster, Donald. *Dust Bowl: The Southern Plains in the 1930s*. New York: Oxford University Press, 1979.

Chapter 15. Community Ecologies

Fitzgerald, Daniel. *Faded Dreams: More Ghost Towns of Kansas*. Lawrence: University Press of Kansas, 1994.
Gilfillan, Merrill. *Magpie Rising: Sketches from the Great Plains*. Boulder, CO: Pruett, 1988.
Kansas Samper Foundation. https://kansassampler.org/.
The Land Institute. https://landinstitute.org/.
Matthews, Anne. *Where the Buffalo Roam: The Storm over the Revolutionary Plan to Restore America's Great Plains*. New York: Grove, 1992.
Nebraska Community Foundation. https://www.nebcommfound.org/.
Payne, James. "Investigation of the Great Plains: Field Notes from Trips in Eastern Colorado." Colorado Experiment Station Bulletin 59, 5–16. Fort Collins: Colorado Experiment Station, 1900.
Penner, Marci, and Wendee Rowe. *The Kansas Guidebook 2 for Explorers*. Newton, KS: Mennonite Press, 2017.
Popper, Deborah, and Frank Popper. "The Great Plains: From Dust to Dust." *Planning* 53 (December 1987): 12–18.
———. "The Buffalo Commons: Its Antecedents and Their Implications." *Online Journal of Rural Research & Policy* 6 (December 2006): 1–26.
Shortridge, James. *Cities on the Plains: The Evolution of Urban Kansas*. Lawrence: University Press of Kansas, 2004.
Wishart, David. *The Last Days of the Rainbelt*. Lincoln: University of Nebraska Press, 2013.

Chapter 16. The Showalter Lilac

Allan, Tom. "Lilac Garden in Full Bloom: Farm near Ogallala Is Home to Four Acres of Fragrant Flowers." *Omaha World Herald*, May 14, 2000.
Dietzel, Ronald. "Lilac Bushes Re-Unite Family." Research report compiled by author, August 2020.

———. "The Good Old Days Were Not That Good (Healthwise)." Research report compiled by author, April 23, 2021.

Harrison, Charles. *The Gold Mine in the Front Yard and How to Work It: Showing How Millions of Dollars Can Be Added to the Prairie Farm.* St. Paul, MN: Webb, 1905.

Kooser, Ted. *Local Wonders: Seasons in the Bohemian Alps.* Lincoln: University of Nebraska Press, 2002.

Lack, Walter. "Lilac and Horse-Chestnut: Discovery and Rediscovery." *Curtis's Botanical Magazine* 17 (May 2000): 109–141.

Peterson, Max. "Meadowlark Hill Lilac Arboretum." *Proceedings of the International Lilac Society Sixteenth Annual Convention* 16 (November 1987): 16.

Chapter 17. Lithophilia

Asher, Brenden. "Across the Central Plains: Clovis and Folsom Land Use and Lithic Procurement." *PaleoAmerica* 2 (April 2016): 124–134.

Bamforth, Douglas. "Paleoindian Perambulations and the Harman Cache." *Plains Anthropologist* 58 (February 2013): 65–82.

Bamforth, Douglas, and Keri Hicks. "Production Skill and Paleoindian Work Group Organization in the Medicine Creek Drainage, Southwestern Nebraska." *Journal of Archaeological Method and Theory* 15 (March 2008): 132–153.

Banks, Larry. *From Mountain Peaks to Alligator Month Stomachs: A Review of Lithic Sources in the Trans-Mississippi South, the Southern Plains, and Adjacent Southwest.* Oklahoma Anthropological Society Memoir 4, 1990.

Greiser, Sally. "Preliminary Statement about Quarrying Activity at Flattop Mesa." *Southwestern Lore* 49 (1983): 6–14.

Hoard, Robert, John Bozell, Steven Holen, Michael Glascock, Hector Neff, and J. Michael Elam. "Source Determination of White River Group Silicates from Two Archeological Sites in the Great Plains." *American Antiquity* 58 (October 1993):698–710.

Hofman, Jack. "Tethered to Stone or Freedom to Move: Folsom Biface Technology in Regional Perspective." In *Multiple Approaches to the Study of Bifacial Technologies*, ed. Marie Soressi and Harold Dibble, 229–249. Philadelphia: University of Pennsylvania Museum of Archaeology and Anthropology, 2003.

Hofman, Jack, and Brendon Asher. "Clovis Activity in the Central Plains Uplands." *Current Research in the Pleistocene* 28 (January 2011): 47–49.

Holen, Steven. "Bison Hunting Territories and Lithic Acquisition among the Pawnee: An Ethnohistoric and Archaeological Study." In *Raw Material Economies among Prehistoric Hunter-Gathers*, ed. Anta Montet-White and Steven

Holen, 399–411. University of Kansas Publications in Anthropology 19. Lawrence: University of Kansas, 1991.

———. "The Age and Taphonomy of Mammoths at Lovewell Reservoir, Jewell County, Kansas, USA." *Quaternary International* 169–170 (July 2007): 51–63.

———. "The Eckles Site, 14JW4: A Clovis Site in Northern Kansas." *Plains Anthropologist* 55 (November 2010): 299–310.

———. "Clovis Lithic Procurement, Caching, and Mobility in the Central Great Plains of North America." In *Clovis Caches: Recent Discoveries and New Research*, ed. Bruce Huckell and David Kilby, 177–200. Albuquerque: University of New Mexico Press, 2014.

Holen, Steven, and Mark Muniz. "A Flattop Chalcedony Biface Cache from Northeast Colorado." *Current Research in the Pleistocene* 22 (2005): 49–50.

Huckell, Bruce, and David Kilby, eds. *Clovis Caches: Recent Discoveries and New Research*. Albuquerque: University of New Mexico Press, 2014.

Michener, James. *Centennial*. New York: Random House, 1974.

Osborn, Alan. "The Baller Biface Cache: A Possible Clovis Site in Hitchcock County, Nebraska." *Plains Anthropologist* 61 (May 2016): 159–176.

Prescott, Cynthia Culver. *Pioneer Mother Monuments: Constructing Cultural Memory*. Norman: University of Oklahoma Press, 2019.

Quigg, Michael, Matthew Boulanger, and Michael Glascock. "Geochemical Characterization of Tecovas and Alibates Source Samples." *Plains Anthropologist* 56 (August 2011): 121–141.

Reher, Charles. "Large Scale Lithic Quarries and Regional Transport Systems on the High Plains of Eastern Wyoming: Spanish Diggings Revisited." In *Raw Material Economies among Prehistoric Hunter-Gathers*, ed. Anta Montet-White and Steven Holen, 251–284. University of Kansas Publications in Anthropology 19. Lawrence: University of Kansas, 1991.

Schroedl, Alan. "The Geographic Origin of Clovis Technology: Insights from Clovis Biface Caches." *Plains Anthropologist* 66 (May 2021): 120–148.

Stanford, Dennis. "Preliminary Report of the Excavation of the Jones-Miller Hell Gap Site, Yuma County, Colorado." *Southwestern Lore* 40 (1974): 29–36.

———. "The 1975 Excavations at the Jones-Miller Site, Yuma County, Colorado." *Southwestern Lore* 41 (1975): 34–38.

———. "Bison Kill by Ice Age Hunters." *National Geographic* 155 (January 1979): 114–121.

Stein, Martin. *Sources of Smoky Hill Silicified Chalk in Northwest Kansas*. Kansas State Historical Society Anthropological Series no. 17. Topeka: Kansas State Historical Society, 2005.

———. "Kansas Lithic Resources." In *Kansas Archaeology*, ed. Robert Hoard and William Banks, 264–282. Lawrence: University Press of Kansas, 2006.

Wedel, Waldo. *Central Plains Prehistory: Holocene Environments and Cultural Change in the Republican River Basin.* Lincoln: University of Nebraska Press, 1986.

Westfall, Tom. *Never a Walk Too Far: Artifact Adventures and Existential Musings.* Sterling, CO: Mammoth Run, 2019.

Wilson, Edward O. *Biophilia: The Human Bond with Other Species.* Cambridge, MA: Harvard University Press, 1984.

Chapter 18. The Greatest Day in the History of Beeler

Anonymous. "'Detour' in Scientist's Life Led Him to a Ness County Sod House." *Salina Journal*, May 22, 1960.

Fuller, Robert, and Merrill Mattes. *The Early Life of George Washington Carver.* Diamond, MO: George Washington Carver National Monument Library, 1957.

Gart, Jason. "He Shall Direct Thy Paths: The Early Life of George Washington Carver." Omaha, NE: Midwest Regional Office, National Park Service, US Department of the Interior, 2014.

Hersey, Mark. *My Work Is that of Conservation: An Environmental Biography of George Washington Carver.* Athens: University of Georgia Press, 2011

Holt, Rackham. *George Washington Carver: An American Biography.* Garden City, NY: Doubleday, Doran, 1944.

Kremer, Gary, ed. *George Washington Carver: In His Own Words*, 2nd ed. Columbia: University of Missouri Press, 2017.

Lennen, O. L. "Carver Sidelights." *Ness City News*, October 15, 1953.

McMurry, Linda. *George Washington Carver: Scientist and Symbol.* New York: Oxford University Press, 1981.

Miner, Craig. *West of Wichita: Settling the High Plains of Kansas, 1865–1890.* Lawrence: University Press of Kansas, 1986.

Shortridge, James. *Kansas City and How It Grew, 1822–2011.* Lawrence: University Press of Kansas, 2012.

Wellman, Paul. "Friends of Old Days in Kansas Saw Budding Genius of Negro Scientist." *Kansas City Times*, September 9, 1942.

Chapter 19. Rock Towns

Aber, Susan, and David Grisafe. *Petrographic Characteristics of Kansas Building Limestones.* Kansas Geological Survey Bulletin 224. Lawrence: Kansas Geological Survey, 1982.

Barbour, Erwin. "Nebraska Green Quartzite an Important Future Industry." *Nebraska Geological Survey Report* 4 (1915): 249–252.
Cutler, Phoebe. *The Public Landscape of the New Deal*. New Haven, CT: Yale University Press, 1985.
Frye, John, and Ada Swineford. "Silicified Rock in the Ogallala Formation." *Kansas Geological Survey Bulletin 64*, pt. 2 (1946): 33–76.
Grisafe, David. *Kansas Building Limestone*. Kansas Geological Survey Mineral Resources Series 4. Lawrence: Kansas Geological Survey, 1976.
Heat-Moon, William Least. *PrairyErth: A Deep Map*. Boston: Houghton Mifflin, 1991.
Muilenberg, Grace, and Ada Swineford. *Land of the Post Rock: Its Origins, History, and People*. Lawrence: University Press of Kansas, 1975.
Risser, Hubert. "Kansas Building Stone." *Kansas Geological Survey Bulletin* 142, pt. 2 (1960): 53–122.
Ruede, Howard. *Sod-House Days: Letters from a Kansas Homesteader*, ed. John Ise. Lawrence: University Press of Kansas, 1983.
Schoewe, Walter. "The Geography of Kansas: Part IV: Economic Geography: Mineral Resources." *Transactions of the Kansas Academy of Science* 61 (Winter 1958): 359–468.
Shortridge, James. *Peopling the Plains: Who Settled Where in Frontier Kansas*. Lawrence: University Press of Kansas, 1995.

Chapter 20. Stickers

Albertson, Frederick, "Man's Disorder of Nature's Design in the Great Plains," Presidential Address to the Kansas Academy of Science. *Transactions of the Kansas Academy of Science* 52 (June 1949): 117–131.
Barton, Erin, Drew Bennett, and William Burnidge. "Holistic Perspectives—Understanding Rancher Experiences with Holistic Resource Management to Bridge the Gap between Rancher and Researcher Perspectives." *Rangelands* 45 (October 2020): 143–150.
Friends of the Kaw. https://kansasriver.org/.
Helzer, Chris. *Hidden Prairie: Photographing Life in One Square Meter*. Iowa City: University of Iowa Press, 2020.
Hill, Dorothy, and Jessie Tompkin. *General and Engineering Geology of the Wray Area, Colorado and Nebraska*. Geological Survey Bulletin 1001. Washington, DC: Government Printing Office, 1953.
Jackson, Wes. *Becoming Native to this Place*. Washington, DC: Counterpoint, 1996.
Paukert, Craig, Joshua Schloesser, Jesse Fischer, Jeff Eitzmann, Kristen Pitts, and

Darren Thornbrugh. "Effects of Instream Sand Dredging of Fish Communities in the Kansas River USA: Current and Historical Perspectives." *Journal of Freshwater Ecology* 23 (December 2008): 623–633.

Prairie Plains Resource Institute. https://www.prairieplains.org/.

Stegner, Wallace. *Where the Bluebird Sings to the Lemonade Springs: Living and Writing in the West*. New York: Random House, 1992.

Whitney, Bill. "A Platte River Country Restoration: Part I. Getting Started." *Restoration & Management Notes* 15 (Summer 1997): 6–15.

———. "Platte River Country Restoration: Part II. At Work on the Plains." *Restoration & Management Notes* 15 (Winter 1997): 126–137.

Epilogue

Dillard, Annie. *Pilgrim at Tinker Creek*. New York: Harper & Row, 1974.

Kooser, Ted. *Local Wonders: Seasons in the Bohemian Alps*. Lincoln: University of Nebraska Press, 2002.

Marsh, Melissa. *Nebraska POW Camps: A History of World War II Prisoners in the Heartland*. Charleston, SC: History Press, 2014.

Index

Note: Color photos have been gathered in a gallery and identified by gallery page number (G1, G2, etc.).

Abilene, KS, xxiv (map), 96, 168, 221, 225
aeolian landforms (loess deposits, sandhills), 8–13, G4 (photo)
African Americans, 215
　and Nicodemus colony, 83, 213
　post–Civil War immigration of, to Kansas City area, 213
agripipers, 118. *See also* shorebirds
Akron, CO, xxiv (map), 244
Albertson, Frederick, 60, 62, 84–85, 231, 241
Alibates agatized chert, 200, 201
Alibates flint quarries, 199
All American Catfish Tournament, 129, 130, 136
Allegawaho Memorial Heritage Park, 33, 35
All the Strange Hours (Eiseley), 150
Alma, KS, xxiv (map), 131, 221
American elm, 38, 40
American golden plover, 117, 120, 123, 166
American redstart, 102
Andrews, Nathan and Laura, 238–241. *See also* Fox Ranch
Anton, CO, xvi, 241
Arapaho (tribe), 48–49, 81, 84, 179
Arapaho, NE, 44
Archidiskodon imperator, 35
Archie (fossil), 35. *See also* mammoth
Arikaree Breaks, 10, 57, G4 (photo)
Arikaree River, xvi, xvii, xviii (map), 3, 4, 8, 11, 46, 86, 134, 137, 177, 197, 201, 207, 239, 241
　Fox Ranch and, 134, 238, 239, G1 (photo)
Argentine Bottoms, xxv (map), 169

Argentine district (Kansas City, KS), xxv (map), 27, 171, 186
Argentine Limestone, 27, 28
Arkansas River, xxiii (map), 4, 8, 21, 62, 78, 80, 81, 84
Armourdale Bottoms, xxv (map), 169
Armourdale district (Kansas City, KS), xxv (map), 171, 186
Art Deco architecture, 229–230
Ash Hollow (NE), 110–111
Ash Rock Church, 225
Assumption of the Blessed Virgin Mary Catholic Church (Dwight, NE), 245
Atlanta, NE, xxiv (map), 243, 246
Atwood, KS, 204
Aurora, NE, xxiv (map), 234–235, 236–237, 238
avocet, American, 116, 119

badger, American, 90
Baird's sandpiper, 117, 120
bald eagle, 166
Bamforth, Douglas, 202–203
Banks' Florilegium, 63
Basilica of Saint Fidelis, 205, 224
basswood, 38, 210
Battle of Punished Woman's Fork, 6
Baumgartner, Fredrick and Marguerite, 106
Bausman family, cemetery lilac of, 189–190
Beatrice, NE, xxiv (map), 29, 44, 223
beaver, American, 163, 165
Beaver Creek (tributary of Republican River), xxiii (map), 3, 77, 82

INDEX

Becoming Native to this Place (Jackson), 231
Beeler, Frank, 211, 212
Beeler, KS, xxiv (map), 209, 211, 212, 215
Belleville, KS, 23
Beloit, KS, xxiv (map), 224
Benedict, Lauryn, 154
Benton, Thomas Hart, 169, 171, 229
Berry, Wendell, xv, 64, 75, 174
Bethany College, 23
Bethany Falls Limestone, 27–28
Big Blue River (Kansas City metro), xix. *See also* Blue River
Big Blue River (tributary of the Kaw), xviii, xix, xxiii (map), 9, 29, 32, 40, 42, 54, 62, 70, 79, 114, 128, 222, 235, 245
big bluestem, 51, 54, 55, 56, 57
Big Creek (tributary of Smoky Hill River), 135–137, 228
Big Creek and its Fishes (Tomelleri et al.), 135–136
big timbers, 45–46, 81–82
Big Timbers of Republican River, 45–46
Big Timbers of Smoky Hill River, 45, 49, 81
Biophilia (Wilson), 196
Bird City, KS, xxiv (map)
    city hall of, 226, 227–228, G8 (photo)
Birds of an Iowa Dooryard (Sherman), 100
Birds of Nebraska (Sharpe et. al), 122
bison, 58–59, 77–87, 88–90, 92, 95
    destruction of, 164–166
    Native American lifeways and, 39, 78–83
Bison antiquus, 197, 198. *See also* Jones-Miller archaeological site
black-and-white warbler, 102
Blackburnian warbler, 102
black-footed ferret, 90, 97–98
Blackmore, William, 164
black-necked stilt, 116
black oak, 38, 40
blackpoll warbler, 102
blackside darter, 131
black tern, 123
bloodroot, 64, 212
blowout grass, 61
blue catfish, 126, 127, 129
blue grama, 51, 57, 58–59, 61, 88, 93, 96, 149, 173
Blue Hills (KS), 19
Blue Mound (KS), 28
Blue Prairie (Kansas City, MO), 50
Blue Rapids, KS, xxiv (map), 32, 147, 221
Blue Ridge (Kansas City, MO), xxv (map), 50–51
Blue River (Kansas City metro), xix–xx, xxi–xxii, xxv (map), 27, 50, 104
bluntnose minnow, 131
Bogue, KS, township hall of, 230
Borland, Hal, 65, 66–67, 155–156
Borthwick George, 214, 216–217
bottoms (bottomlands), Kansas City area, xxv (map). *See also* Argentine Bottoms; Armourdale Bottoms; East Bottoms; Rosedale Bottoms; West Bottoms
Bourgmont, Etienne de, 79, 80
Bowersock Dam, 127, 128
branchiopods (freshwater shrimp), 156, 158
brassy minnow, 133, 134
Brooking, Albert Munsell, 108
Brooks, David, 185
Brookville, KS, 223
Brown, Corie, 176, 177, 179
brown creeper, 103
Brulé band of Lakota (tribe), 39, 81, 83
Bryan, Edwin, 51
Buehler, Dawn, 232–234, 241. *See also* Friends of the Kaw; Kansas Riverkeeper
buffalo. *See* bison
Buffalo Commons
    Capital of the, 182 (*see also* McCook, NE)
    proposal of Frank and Deborah Popper, 181–182
    Storytelling & Music Festival, 182 (*see also* McCook, NE)
buffalograss, 51, 57, 58–59, 88, 90, 149, 173
Buffalo Hunters, The (Sandoz), 77, 78
buff-breasted sandpiper, 117–119, 120, 123, 124, 166
Bullock's oriole, 107, 108
Burlington, CO, xxiv (map), 9
bur oak, 38, 40, 41–43, 65
Bur Oak Canyon, 41
bur oak woodland, 40, 41
burrowing owl, western, 91–92, 97
bush morning glory, 73
butterfly milkweed, 143

Camp Atlanta, 244
Camp Funston, 221
cardinal, northern, 100, 111, 120
Carlisle Shale, 19
carmine shiner, 131, G7 (photo)
Carolina parakeet, xv–xvi, xviii
Carson, Rachel, 12
Carver, George Washington, 209–217
  homestead site of, in Ness County, KS, 209, 211–212, 214, 216, 217
Cassin's sparrow, 60, 105, 106
Castaner, David, xxi
Catharine, KS, 225
Cathedral of the Plains, 224. *See also* Basilica of Saint Fidelis
Cather, Willa, 22–23, 37, 55–57, 107, 161, 173
cattle, 54, 59, 90
  Black Angus breed of, 90, 240
cedar, eastern red, 22, 47
Cedar Point (CO), xxiii (map), 4–5, 8, 36, 47–48
cedar waxwing, 111, 112
Centennial (Michener), 196, 203
Central Missouri State University, xx–xxi
Central Plains Prehistory (Wedel), 196
channel catfish, 126, 129, 130
Chase County (KS) Courthouse, 218, 219
chert, 29, 199. *See also* Alibates agatized chert; Flint Hills; Hartville chert; *see under* Ogallala Formation
Chester, NE, 181
chestnut oak, 38–39, 40
chestnut-sided warbler, 102, 103
Cheyenne (tribe), 6, 46, 81, 83, 84, 86, 179
  attack on encampment of, at Sand Creek, 48–49
  bison and lifeways of, 81–82
  winter camps of, in big timbers, 81–82
Cheyenne Bottoms, 115, 117
Cheyenne Bottoms (Zimmerman), 113
Cheyenne Wells, CO, xxiv (map), 176
Chinese elm, 177, 178. *See also* Siberian elm
Chisolm Trail, 95–96, 168
Chivington, John, attack led by, against Cheyenne and Arapahoe at Sand Creek, 48–49
Chokecherry Places (Gilfillan), 100

Christ Episcopal Cathedral Church, 222
Christ with Lamb (Felton), 205
Cities on the Plains (Shortridge), 187
City of Native Stone, 221. *See also* Alma, KS
Clark, William, xv–xvi, xviii, 27
Clay Center, KS, 221
Clay County (KS) Courthouse, 221, 229
Clean Water Act of 1972, 129
Clements, Fredrick, 56, 62
climate and weather
  bomb cyclone, 115
  derecho, 42
  drought, 52, 84, 132, 172–174, 180, 189, 183, 213, 222, 227
  influence of, on grassland formations, 51–52
  wind, 8–13, 52
  See also Dust Bowl
Cloud County (KS) Museum, 223
Cloverly quartzite, 199. *See also* Spanish Diggings
Clovis people, 36, 196, 197, 198, 201, 202, 203, 204
Colby, KS, xxiv (map)
  city hall of, 225, 229–230
Colorado Gold Rush, 21, 84
Colorado Piedmont, xxiii (map), 4, 5, 47, 200
Colorado State University, 176, 183
common shiner, 131, G7 (photo)
common yellowthroat, 102
community disassembly, 158
Concordia, KS, xxiv (map), 17, 18, 223
Conzen, Michael, 177
Cope, CO, xvi, xxiv (map)
Coronado, Francisco Vasquez de, 17, 80–81
Coronado Heights, 17, 24, G2 (photo)
Costello, David, 58, 75, 133
cottonwood, eastern, 38, 40, 43–47, 49, 55, 81
  as winter fodder for horses, 46–47
Cottonwood Limestone, xxi, 29, 218–219, 221, 222
Cottonwood Ranch State Historic Site, 226
cottonwood riparian woodland, 40, 45–46, 86, 107, G1 (photo). *See also* big timbers
Country Editor's Boy (Borland), 155
cowbird, brown-headed, 90
cowboy's delight, 94. *See also* scarlet globe-mallow

cowtown, 168. *See also* Abilene, KS; Ellsworth, KS; Kansas City
creek chub, 131, 133
Cretaceous Period, 15, 21, 34, 35
Crossing the Kansas (Miller), 125
cuestas, description of, 26, G2 (photo)
Culbertson, NE, 57
Custer, George Armstrong, 46, 83, 222

Dakota Aquifer, 17–18, 19, 123
Dakota Formation, 14, 15, 16–17, 223
 fossils discovered in, 34
 use of, sandstone as structural stone, 17, 223, 228
Dakota Hills, 16–19, 23–24, G2 (photo)
Dakotanthus, 34
Damar, KS, 225
Darwin, Charles, 139, 145
Davis, W. W. H., 52–53, 94
Deer Creek (NE), 10
Delaware (tribe), 79, 81, 82–83
Delaware Outlet, 82–83
Delaware River, xviii, 32, 128, G7 (photo)
Denver, CO, xx, xxi, xxiii (map), 77, 182
Denver Museum of Nature & Science, 200, 201
Depression-era, 17, 222, 225, 226, 230. *See also* Great Depression
De Soto, KS, 233
Dietzel, Ron, 194, 195
Dillard, Annie, 158, 243
disturbance, ecological, 88–90
disturbance agents, 88–89, 90, 95–96
Dodge, Richard Irving, 78, 164
dog nomads, 80, 179
Dorchester, NE, 181
Douglas County (KS) Courthouse, 221, 229
Downs, KS, 16, 176
downy yellow violet, 66
dredging industry, sand and gravel, 128, 233–234
Driftwood Creek, 41
drought, 132, 173–174, 180, 183, 213, 222
 great, of 1930s, 52, 172–174, 180, 227
 See also Dust Bowl
Dull Knife, Chief (Cheyenne), 6
Dunbar, John, 80

dung beetles, 90
dunlin, 116, 117, 124
Dust Bowl
 drought, 133, 172–174, 180, 227
 geographical extent of, 173
 Great Depression and, 172
 timeframe of, 108, 172, 227
dust storms, 173
Dutchman's breeches, 64, 140
dwarf milkweed, 89, 142–143
Dwight, NE, xxiv (map), 245, 246
Dyck Arboretum of the Plains, xxi

eagle, skeleton of, buried in grave of Pawnee child, 111. *See also* bald eagle; golden eagle
East Bottoms (Kansas City, MO), xix, xxv (map)
Easter daisy, 65, 66
eastern deciduous forest, 37–38, 39, 40, 64, 86, 102
eastern kingbird, 100, 107
eastern meadowlark, 105
Easton, L. C., 86–87
Eckles, Dick, 203
ecological bottleneck, 46
ecological cascade, 89, 93
ecological partitioning, 102, 131–132
"Ecology of Mixed Prairie in West Central Kansas" (Albertson), 85
ecosystem engineers, 88, 90–95
Edge of the Sea, The (Carson), 12
Egan, Timothy, 173
Eiseley, Loren, 150, 159
Eisenhower, Dwight D., 204, 221, 232
El Cuartelejo (also rendered El Quartelejo), 6
elk, 30, 79
Elkader, KS, xxiv (map), 88, 96, 98
Ellis County (KS) Courthouse, 225, 227
Ellis, KS, 136, 225
Ellsworth, KS, xxiv (map), 34, 96, 164, 168, 215, 223
Elmo, KS, 34
Emanuel Lutheran Church, 225
"Enchanted Bluff, The" (Cather), 22
Encyclopedia of the Great Plains (Wishart), 177

Enders Reservoir State Recreation Area, 153
ephemeral habitat, 133, 156. *See also* playas
Epler, Stephen, 181
Eskimo curlew, 166–168
"Eskimo Curlew and Its Disappearance" (Swenk), 167
Essay on the Geography of Plants (Humboldt), 1
Eurasian skylark, 104–105
evening primrose sweat bee, 146–147
Exodusters, African-American, 83, 213, 225. *See also* Nicodemus, KS
Exposition Internationale des Arts Decoratifs et Industriels Modernes, 228–229. *See also* ArtDeco architecture
Extermination of the America Bison (Hornaday), 164
extinction
  of Carolina parakeet, xv–xvi
  of Eskimo curlew, 166–168

Fairbury, NE, xxiv (map), 17, 33, 228
Fairview Cemetery, 189, 190, 191, 193, 194, 195
fairy shrimp, 156, 157, 158
fathead minnow, 133
felsenmeer, 31, 33. *See also* glacial erratics
Felton, Pete, 204–207, 232
Fencepost Limestone, 19–20, 223–224
  use of, in fencing, 19–20, 207, 214, 223, G3 (photo)
  use of, in sculpture, 206, 208
  use of, as structural stone, 19, 20, 24, 219, 223–225, 227, 230
Fenneman, Nevin, 3, 4
ferruginous hawk, 90, 97, 98, 109
"Field Notes from Trips in Eastern Colorado" (Payne), 177
Filley Stone Barn, 222–223
fire, influence of, on plant communities, 39, 47, 52, 53
fishes
  in Flint Hills streams, 29–30, 130–132, G7 (photo) (*see also* Fox Creek; Kings Creek; Mill Creek)
  in High Plains streams, 132–137(*see also* Arikaree River; Ogallala minnows)
  in main stem of Kaw River, 126–130

Fish-within-a-Fish (fossil), 34–35
Fish Wizards Tour, 131
Fitch, Julian, 77, 79, 86
flathead catfish, 126, 129, 130
Flattop Butte, 199, 200, 201
Flattop chalcedony, 200, 201, 202
flint, 29, 199. *See also* chert
Flint Hills, xxiii (map), 25–26, 28–29, 30, 40, 65, 123, 244, G2 (photo)
flintknapping, 198, 203, 204. *See also under* Paleoindian
Flood Disaster (Homecoming—Kaw Valley) (Benton), 171, 229
flooding
  influence of, on cottonwood ecology, 44–45
  influence of, on Kansas City, KS, bottomland neighborhoods, 171, 186
  influence of, on riverine habitat structure, 127–128
floodplain forest, 40, 43
"Floral Clock for Kansas, A" (Smythe), 145
Florence Limestone, 29
Flores, Dan, 78, 84, 87, 94, 164
Folsom people, 197, 198, 201
football, six-man, 181
forest. *See* floodplain forest; gallery forest; oak-hickory forest
Forming Camp (Miller), 125
Fort Fletcher Stone Arch Bridge, 228
Fort Hays Limestone, 19, 20–21, 24, 57, 206, 225
  use of, as structural stone, 24, 225
Fort Hays State University, 84, 135, 136, 205, 225, 231
Fort Riley, KS, 222
Fort Riley Limestone, 29, 206, 221–223
Fort Riley Military Reservation, 221, 222
Fort Scott, KS, 210, 216
fossils, discovery of, in Kaw watershed, 33–35
Four Famous Kansans (Felton), 204–205, 206, 232
Fox Creek, 131, 132. *See also* Tallgrass Prairie National Preserve
Fox Ranch, xxiv (map), 134, 238–241, G1 (photo). *See also* Arikaree River
Franklin's gull, 120–123, 124, G6 (photo)

Frazier, George, 97
Frémont, John Charles, 42, 51, 58, 63, 69–72, 77, 79, 82
Fremont's clematis, 71, 72, G5 (photo)
Fremont's evening primrose, 71, 212, G5 (photo)
Frenchman Creek, xxiii (map), 3, 7, 11, 57
Friends of the Kaw, xviii, 232–234
Funston Limestone, 221

gallery forest, Flint Hills, 40, 65, 130
Geary County (KS) Courthouse, 222
Geodetic Datum for the United States, 15
Geographical Center of the United States, 15–16
Geosternbergia sternbergi, 35
Geyer, Charles Andreas, 68, 69
Gido, Keith, 126, 131, 135
Gilfillan, Merrill, 100, 111, 177–178
Gilmore, Melvin, 138
glacial erratics, 31, 32–33
Glacial Hills, 31–32, xxiii (map), 31–32
golden eagle, 8, 90, 97, 109–110
goldfinch, American, 100, 105
Gold Mine in the Front Yard and How to Work It, The (Harrison), 191
Gorham, KS, 224
Gothic architecture, 224
Gothic Revival architecture, 222
Graham County (KS) Public Library, 204, 206
Grande Riviere des Cansez, xvii
Grant, NE, 191
grasshoppers, outbreaks/plagues of, 108, 213, 222
Grasslands of the Great Plains (Weaver and Albertson), 60
Great Depression, 227
  Dust Bowl and, 227
  Works Progress Administration and, 17, 33
greater prairie chicken, 239, 240
greater yellowlegs, 116, 117
"Great Plains: Dust to Dust" (Frank and Deborah Popper), 181–182
Great Platte River Road, The (Mattes), 67

Great Plow-Up, 174. *See also* Dust Bowl
Greenhorn Limestone, xxi, 14, 15, 19–20, 57, 223–224
Greeley, Horace, 220
green ash, 38, 40
Grinnell, George Bird, 86, 87
Grinnell, KS, 135
Griset, Ernest, 164
Guide Rock, NE, xxiv (map), 22, 73, 80
Guide Rock, The, 22

hackberry, 38, 40
Hackberry Creek (tributary of Smoky Hill River), 3
Harrison, Reverend Charles Simmons, 69, 191
Harris's sparrow, 103–104
Hartville chert, 199. *See also* Spanish Diggings
Hartwell, Don (and Vera), 173
harvester ants, 93
hawk moths, 146, 149. *See also* Missouri evening primrose; white-lined sphinx
Hayes Center, NE, 182
Hays, KS, xxiv (map), 41, 82, 84, 85, 204, 205, 224, 225, 227
Heat-Moon, William Least, xvii–xviii, 218, 219
Hebron, NE, 181
hedge bindweed, 145
Hell Gap people, 198. *See also* Jones-Miller archaeological site
Helzer, Chris, 237
hermit thrush, 103
Herron, John, 168
Hesperornis regalis, 35
Hesston, KS, xx, xxi, xxiii (map), xxiv (map), 53, 101, 104, 189, 194
Hidden Prairie (Helzer), 237
Highland College, 211. *See also* Carver, George Washington
High Plains, xxiii (map), 3–13, G4 (photo)
  aeolian landforms (loess deposits, sandhills) of, 8–13, G4 (photo)
  erosion landforms (bluffs, buttes, canyons, escarpments) of, 5–7
  Ogallala Formation and, 4–7

INDEX

High, Wide, and Lonesome (Borland), 65, 67, 155
Hi-Line, 13
Hill, Asa T., 111
Hill City, KS, xxiv (map), 24, 204, 206–207, 215, 225, 226, 228, 231
Holdrege, NE, xxiv (map), 41
Holen, Steven, 201
Holistic Management grazing, 239
Holy Cross Catholic Church, 224
Homecoming – Kaw Valley 1951 (Benton), 171
Hooker's sandwort, 151, 152
Hopkins, Gerard Manley, 105, 107
Hornaday, William, 164
horned larks, 92–93, 105, 106–107, 109
horse nomads, 39, 46, 81, 85, 179
horses, 46–47, 92
Hudsonian godwit, 116, 119–120, 166, 246
Humboldt, Alexander von, 1
Hunting and Trading on the Great Plains (Mead), 165
Hunting Grounds of the Great West (Dodge), 78, 164
hunting, market, xv, 89, 94, 167

Ice Age Monument, 32
Idalia, CO, 240
Immense Journey, The (Eiseley), 150
Imperial Dune Field, 11, 12
Inavale, NE, 173
Independence, MO, xxv (map), 28, 50, 52, 67, 104
Indiangrass, 51, 54, 55
Indians. *See* Native Americans
intermediate wheatgrass, 184
International Lilac Society, 192–193
Interstate Highway 70, xx, 82, 135
Íⁿ'zhúje 'waxóbe, 33, 36. *See also* Sacred Red Rock; Shunganunga boulder

Jack-in-the-pulpit, 64
Jackson, Dana, 184
Jackson, Wes, 51, 184, 231, 242
James, Edwin, 43, 92, 94, 95
Jamestown Marsh, 18
Janovy, John, Jr., 153

Jefferson City, MO, xxi, 169
Jefferson, Thomas, 234
Jewell County (KS) Courthouse, 225, 227, 229, G8 (photo)
Johnny darter, 131
Johnsgard, Paul, 91, 121
Jones, Janie (and Robert), 207, 208. *See also* Jones-Miller archaeology site
Jones-Miller archaeology site, 197, 198, 199, 200–201, 207–208
Jorgensen, Joel, 115, 118, 124
Junction City, KS, xviii, 129, 222

Kanopolis Reservoir, 17, 128
Kansa (tribe), xvii, 33, 78, 79, 81, 84, 179
  communal bison hunt of, 80–81
  reservation of, in Kansas, 82–83
  See also Kaw Nation
Kansas Academy of Science, 95, 145, 241
Kansas and the Country Beyond (Copley), 14
Kansas antelope sage, 6
Kansas City and How it Grew (Shortridge), 55, 168–169
Kansas City (general area)
  bottomlands in, xxv (map), 129, 168–171, 186, 187
  flooding of Kaw River in, 127, 171, 186, 229
  forest, woodland, and savanna in, 37–39, 40
  as gateway city, 177
  meatpacking industry of, 26, 168–171
  stockyards, 26, 168, 169, 170, 171
  tallgrass prairie in, xxv (map), 39, 50–51, 55
Kansas City, KS, xxiii (map), xxiv (map), xxv (map), 26, 27, 129, 171, 186, 187, 218, 238
Kansas City, MO, xviii, xx, xxiii (map), xxiv (map), xxv (map), 26–27, 53, 55, 230, G1 (photo)
  Pollution Control Department of, xxi
Kansas City Power and Light Building, xx, 229, 230, G1 (photo)
Kansas Fishes (Kansas Fishes Committee), 136
Kansas Guidebook (Penner and Rowe), 185–186, 219
"Kansas Mountains," 25, 30. *See also* Flint Hills

Kansas Pacific Railroad, 164, 168
Kansas Place-Names (Rydjord), 98
Kansas Post Rock Limestone Coalition, 20
Kansas River, xvii, xviii, 84, 126. *See also* Kaw River
Kansas Riverkeeper, 232. *See also* Dawn Buehler
Kansas Riverkings, 122
Kansas River National Water Trail, 232
Kansas Sampler Foundation, 185–186, 219
Kansas State Capitol, 204, 218, 232
Kansas State University, 183, 221
   Fish Ecology Lab, 126, 131
   Konza Prairie Biological Station and, 54, 130
Kansas Water Authority, 233
Kansas Wild Flowers (Stevens), 69, 138, 145
Kanza, xvii, xviii, 78. *See also* Kansa (tribe)
Kaw Nation (tribe), xviii, 33. *See also* Kansa (tribe); Sacred Red Rock
Kaw Point, xviii, xix, xxiii (map), xxv (map), 4, 27, 32, 51, 74, 78, 129, 187, 229, G1 (photo)
Kaw River, xxiii (map), xxv (map), 126–130, 232–234, G1 (photo), G7 (photo)
   bottomlands of, in Kansas City metro, xxv (map), 168–171
   confluence of, with Missouri River, xv, xviii, xxv (map), 27, 32, 37, 43, 55, 168, 187, 213, 232, G1 (photo) (*see also* Kaw Point; Kawsmouth)
   ecological impact of sand/gravel dredging in, 128, 233–234
   flood of 1844, 127
   flood of 1903, 127, 186
   flood of 1951, 127, 171, 186, 229
   as place name, xvii–xviii
   pollution of, 128–129, 168–170, 233
   recreational paddling on, 126, 127, 129, 232, 234
Kawsmouth, xvi, xviii
Kernza®, 184. *See also* The Land Institute
Kerouac, Jack, 180
kettle of hawks, xvi–xvii, 243–244. *See also* Swainson's hawk
keystone ecology, 89, 90, 99
Keystone Gallery, 98

Keystone, KS, 98–99
Kill Creek Prairie, 54
killdeer, 93, 117, 156
killifish, northern plains, 133, 136
Kindscher, Kelly, 67–68, 73–74
kingbird town, 177–179
Kings Creek, 130–131, 132, 137. *See also* Konza Prairie Biological Station
Kira ru tah (Pawnee: "Manure River"), 87. *See also* Republican River
Kolesnikov, Leonid, 193
Konza Prairie (Reichman), 126
Konza Prairie Biological Station, xxiv (map), 54, 130, G2 (photo)
Kooser, Ted (and Kathleen), 188, 245

Ladder Creek, 4, 5, 6
Lake Scott State Park, xxiv (map), 6, 153
Lakota (tribe), 9–10, 39, 81, 83, 111, 138, 179. *See also* Brulé band of Lakota; Oglala band of Lakota
Land Institute, The, 183–184
Land of the Post Rock (Muilenburg and Swineford), 20
Landscape of Hemispheric Importance to Shorebirds, 116, 118, 123. *See also* Flint Hills; Rainwater Basin
lark buntings, 93, 105–106, 107
Last Child in the Woods (Louv), xix
Last Days of the Rainbelt (Wishart), 96, 179
Last Wild Places of Kansas, The (Frazier), 97
Lauritzen Gardens, xxi
Lawrence, KS, xviii, xxiv (map), 28, 33, 54, 125, 127, 186, 221, 234
   Sacred Red Rock and, 33, 35–36
least sandpiper, 117
least tern, 127
Lebanon, KS, 15, 16
Lebold Mansion, 225
lemon scurfpea, 61
Leopold, Aldo, 142
Leoti, KS, 225
lesser yellowlegs, 117
Lewis, Meriwether, 27, 234
Lewis and Clark expedition, xv, 18, 27, 78, 234
Liebenthal, KS, 224
Life is a Miracle (Berry), xv

Life of a Fossil Hunter (Charles Sternberg), 34
lilacs, 188–195
Lincoln, KS, xxiv (map), 224
Lincoln, NE, xx, xxi, xxiii (map), xxiv (map), 32, 53, 56, 116, 206, 235
Lincoln County (KS) Courthouse, 224, 229
Lincoln County Dune Field (NE), 11, 12
Lincoln Creek (NE), 235, 236, 237, 238
Lindsborg, KS, xxiv (map), 17, 23–24, 228
lithic scatter, 197, 204
lithophilia, 196, 207, 208
Little Blue River (MO), 50
Little Blue River (tributary of Big Blue River), xxiii (map), 9, 17, 42, 54, 110, 114
little bluestem, 20, 51, 52, 54, 56–58, 217
Little Grand Canyon, 10
Little Jerusalem Badlands State Park, 21–22, G3 (photo)
Local Wonders (Kooser), 188, 245
loess, deposition and landforms of, 9–11, 12, G4 (photo)
Loess Canyons (NE), 9–10, 35, 47, 57
Loess Plain, xxiii (map), 9. *See also* Rainwater Basin
Long, Stephen, Expedition of, 43, 92
long-billed dowitcher, 116, 117
Louv, Richard, xix
Love Song to the Plains (Sandoz), 161
lowland prairie, 54–55
Lucas, KS, 186

Magoffin, Susan Shelby, 50, 51
Magpie Rising (Gilfillan), 178
Mallea, Amahia, 169
mammoth, 35, 198, 203
Manhattan, KS, xix, xxiv (map), 221
Mankato, KS, xxiv (map), 225, 227, 229
"Man's Disorder of Nature's Design in the Great Plains" (Albertson), 241
"Manure River," 77, 87. *See also* Republican River
marbled godwit, 116, 119
Massacre Canyon, 9–10, 83
Mattes, Merrill, 67
Matthews, Anne, 182
Maupin, Tom, 42–43
Mayapple, 64, 212

McCarty, John, 118, 124
McCook, NE, xxiv (map), 182
McNish Park (Fairbury, NE), 33, 228
Mead, James R., 95–96, 163, 164, 165–166
Meadowlark Hill Lilac Farm, 191, 193
Mead's milkweed, 143
meatpacking industry, Kansas City, 26, 168–171
Medicine Creek (NE), 10, 13, 244
megafauna, hunting of, by Clovis people, 198, 204
Meganeuropsis permiana, 34
Memoirs of My Life (Frémont), 63
merlin, 109
Michener, James, 196, 203
milkweeds, 89, 130, 142–145
Mill Creek (Johnson County, KS), xix
Mill Creek (Wabaunsee County, KS), 79, 130, 131, 132, 137, 221
Miller, Alfred Jacob, 125
Miner, Craig, 209, 215
Minneapolis, KS, xxiv (map), 18, 211
Missouri evening primrose, 71, 146–147, 149
Missouri River, xvii, xxiii (map), xxv (map), 129
confluence of Kaw River with, xv, xviii, xxv (map), 27, 32, 37, 43, 55, 168, 187, 213, 232, G1 (photo)
Missouri Water Pollution Board, xxi, xxii
Mitchell County (KS) Courthouse, 224, 229
mixed-grass prairie, 20, 51, 52, 55–57, 62, 69–72, 212, G3 (photo)
mockingbird, northern, 108
Morris, Lee, 118
mosasaur, 35
"Mother Kaw," 125, 126, 130
mountain plover, 92, 97
Mount Oread, 28, 221
Mount Sunflower, xxiv (map), 7–8
Muilenburg, Grace, 20
Munjor, KS, 205, 224
Mushroom Rocks State Park, 18
mutualism
    ant/plant, 149 (*see also* myrmecochory; Nuttall's violet)
    grazing, 59, 95
    obligate, 147 (*see also* plains yucca; yucca moth)

My Ántonia (Cather), 22, 37, 56
myrmecochory, 149. *See also* ant/plant; Nuttall's violet

Naponee, NE, 173
narrow-leaf purple coneflower, 71, 141, 212
Nashville warbler, 103
National Register of Historic Places, 219, 223, 228, 230
Native Americans
  dispossession of, 179
  lifeways of, and bison, 78–83
  See also Arapahoe; Cheyenne; Delaware; Kansa (Kaw); Lakota; Omaha; Osage; Otoe; Pawnee; Plains Apache; Potawatomi; Pueblo Indians; Shawnee; Wichita
"Native Hill, A" (Berry), 174
Natural Systems Agriculture, 184. *See also* The Land Institute
Nature Conservancy, The (TNC), 236, 237, 238
  Fox Ranch and, 134, 238–240, photo (G1)
  Konza Prairie Biological Station and, 54, 130, photo (G2)
  Smoky Valley Ranch and, 98, 225
  Sustainable Rivers Program of, 128
  Tallgrass Prairie National Preserve and, 54
Nebraska Community Foundation, 184–185
Nebraska Sandhills, 11, 60, 119, 145
Nebraska Statewide Arboretum, xxi
needle-and-thread (grass), 57, 61
Ness City, KS, 214, 216, 217
Ness County News, 212, 213, 215
New Almelo, KS, 226
New Deal programs, projects supported by, 227–230. *See also* Public Works Administration; Works Progress Administration
New York Times, 182, 185
Nicodemus, KS, xxiv (map), 83, 213, 225
Niobrara Formation, 14, 15, 20–23, 200, 206, 222, 225
No Little People (Schaeffer), 176
Norton, KS, 227
Norton County Historical Museum, 227–228

nuthatch, white-breasted, 100, 103
Nuttall, Thomas, 104
Nuttall's violet, 65, 66, 67, 149

oak-hickory forest, xix, 37, 38, 107
Oakley, KS, xxiv (map), 22, 98, 225
Oberlin, KS, xxiv (map), 204, 206
Ogallala Aquifer
  depletion of, 133–135, 239
  groundwater recharge of, 7
  as source of springs and streamflow, 5, 86, 132–135,
  sustainable management of, 135
Ogallala Formation, 4, 5–7
  agatized chert, 226
  Ogallala Aquifer and, 5, 132
  use of, as structural stone, 225–226
Ogallala Formation quartzite
  use of, as structural stone, 214, 226, 227, 228, 230
Ogallala minnows, 133, 134, 135. *See also* fishes
Oglala band of Lakota (tribe), 81, 83
O. K. Creek, xxv (map), 170
Oklahoma Bird Life (Baumgartner and Baumgartner), 106
Olathe, KS, 54, 69, 210
Omaha (tribe), last bison hunt of, 83
Onaga, KS, xxiv (map), 221, 244
Onaga limestone, 221. *See also* Funston Limestone
O Pioneers! (Cather), 22, 56
On the Road (Kerouac), 180
orange-crowned warbler, 102
Oread Limestone, 28, 221. *See also* Blue Mound; Mount Oread
Oregon Trail, 50, 67, 79, 104, 110, 222
Origin of Species, The (Darwin), 145
Osage (tribe), 79, 81
Osage Cuestas, 17, 26–28, 29, 31, 36
  geology and landscape character of, 25–26
  regional setting of, xxiii (map)
  See also Argentine Limestone; Bethany Falls Limestone; Oread Limestone
Osborne County (KS) Courthouse, 225, 229
Osborne, KS, 225
Otoe (tribe), bur oak and, 42

Our Lady of Perpetual Help Catholic Church, 223
oval-leaf bladderpod, 151
ovenbird, 102

Padani wakpah (Lakota: "Pawnee River"), 81. *See also* Republican River
paddlefish, 126, 128
Pa:hu:ru' (Pawnee: "the rock that points the way"), 22, 80. *See also* Guide Rock, The
Paine, Robert, 99
Paleoindian
    cultural complexes, 197–198
    lithic caches, 197, 201–202
    lithic resources, 197, 200–202 (*see* Alibates agatized chert; Cloverly quartzite; Flattop chalcedony; Hartville chert; Smoky Hills silicified chalk)
    lithic transport, 197, 200–202
    projectile point manufacturing technology, 198–199, 202–203
    quarry sites, 199–201, 204 (*see* Alibates; Flattop Butte; Smoky Hills; Spanish Diggings)
    See also Clovis people; Folsom people; Hell Gap people
pale purple coneflower, 141
pallid sturgeon, 126
palm warbler, 103
Pampas grassland of South America, xvi, 108, 113, 123
Pangaea, 25–26
Paradise, KS, xxiv (map), 163, 228
Paradise Creek, 42, 163, 164
Pattison, Nathan and Rachel, 110, 111
Pawnee (tribe), 9, 22, 45, 79, 111, 179
    attack of Lakota on, in Massacre Canyon, 9–10, 83
    bison and lifeways of, 22, 46, 79–81, 83
Pawnee Buttes, 200, 203
Pawnee Trail, 22, 80, 82. *See also* Guide Rock
Payne, James, 176–177, 179, 180
pectoral sandpiper, 116, 117
Pendergast, Tom, 169
Penner, Marci, 185. *See also* Kansas Sampler Foundation
Pennsylvanian Period, 26, 27

peregrine falcon, 109, 124
Perkins Table, 13, 191
Permian Period, 26, 29, 34, 218
Peterson, Max (and Darlene), 191, 192, 193. *See also* Meadowlark Hill Lilac Farm
petroglyphs, sites of, in Dakota Hills, 18
Pfeifer, KS, 224
Phillipsburg, KS, 227
Pickwell, Gayle, 106, 107
Pike, Zebulon, 3, 30, 79
Pikes Peak, 8
Pilgrim at Tinker Creek (Dillard), 243
Pioneer Family (Felton; Oberlin, KS), 204, 206
Pioneer Family (Felton; Victoria, KS), 205–206
Pioneer Mother Monuments (Prescott), 206
piping plover, 127
Plains Apache (tribe), 79
plains darter, 131, 133
plains minnow, 128
plains stoneroller, 131, 133
Platte River, xvii, xxiii (map), 21, 22, 62, 79, 80, 84, 86, 87, 103, 110, 115, 236, 237, 245
Playa Lakes Joint Venture, 7
playas, 7, 87, 151, 155–158, G4 (photo)
    biodiversity associated with, 7, 86, 87, 123, 156, 157, 158
Pleistocene epoch, 35, 48
plesiosaur, 35
Plum Creek (tributary of Big Blue River, NE), 245
pollination, 138–139
    Dutchman's breeches, 140
    Missouri evening primrose, 146–147
    plains yucca, 147–148
    woodland herbs, 64, 65, 139–140
ponderosa pine, 47–48
Popper, Frank and Deborah, Buffalo Commons proposal of, 181–182, 183
post oak, 38, 40
post rock, 20, 206. *See also* Fencepost Limestone
Post Rock Country, 20, 72, 224, G3 (photo)
Potawatomi (tribe), 79, 83
Potawatomi County (KS) Courthouse, 220
prairie, 50–62. *See also* lowland prairie; mixed-grass prairie; sandsage prairie; shortgrass prairie; tallgrass prairie

Prairie Center (Johnson County, KS), 54
prairie cordgrass, 54
Prairie Dog Creek (tributary of Republican River), 3, 230
prairie dogs, black-tailed, 90–99
  biodiversity associated with, colonies, 90–93, 95
  eradication efforts, 96–97
  historical extent of, colonies in Kaw River watershed, 95–97
  impact of, eradication efforts on associated wildlife, 90, 97, 98
"Prairie Dog Situation in Kansas, The" (Scheffer), 88, 97
prairie dove, 122, 123. *See also* Franklin's gull
prairie falcon, 90, 93, 109
"prairie madness," 13
prairie merlin, 109
prairie phlox, 69, 70
Prairie Plains Resource Institute, 235–238, 241
prairie sandreed, 61, 73
Prairie Smoke (Gilmore), 138
prairie violet, 141–142
Prairie World, The (Costello), 75
PrairyErth (Heat-Moon), xvii–xviii, 218
Prescott, Cynthia Culver, 206
pronghorns, 30, 79, 92, 93–95, 244
Pruett, Craig, 126, 129
Pteranodon sternbergi, 35
pterosaur, 35
Public Works Administration, 227
Pueblo Indians, escape of, to El Cuartelejo, 6
Punished Woman's Fork (tributary of Ladder Creek), 6
purple coneflower, 140–142. *See also* narrow-leaf purple coneflower; pale purple coneflower; Topeka purple coneflower
purple milkweed, 143

Quality Hill, xxv (map), 26–27
Quivira, 17

Rainbelt, 180
Rainwater Basin, xxiii (map), 9, 113–124, 181, 243, 246, G6 (photo)
  seabirds and, 120–123
  shorebirds and, 9, 114–120, 166–167

Rainwater Basin Joint Venture, 124
Ramaley, Francis, 72
rattlesnakes, prairie, 91, 113
ravens, common, 8, 89, 166
Rawlins County (KS), Courthouse, 204
Red Cloud, NE, xxiv (map), 22, 32, 56, 57, 173, 185
redbud, 38, 40
red oak, 38, 40
red shiner, 133
red-tailed hawk, xvi
regal fritillary, 141–142
Reichman, O. J., 126
Republican [bison] herd, 78
"Republican country," 78
Republican River, xxiii (map), 7, 9–11, 27, 59, 70, 84, 87, 177, 196, 182, 201
  confluence of, with Smoky Hill River, xviii, xxiii (map), 40, 232
  dune fields in watershed of, 11–12, 59
  flood of 1935, 127
  headwater streams of, 3, 7 (*see also* Arikaree River; Beaver Creek; Frenchman Creek; North Fork; Prairie Dog Creek; Sappa Creek; South Fork)
  North Fork of, 3, 7, 177
  South Fork of, 3, 86, 110, 143
Richardsonian Romanesque architecture, 229
Riley County (KS) Courthouse, 221, 229
River in the City of Fountains, A (Mallea), 169
Rivers & Birds (Gilfillan), 111
Rockefeller Prairie, 54
Rock City, 18, 212
Rock Island Bridge, 187
rock wrens, 153–155
Rocky Mountain juniper, 47–48
Rocky Mountains
  orographic winds and, 8–9
  rainshadow of, 52
Roosevelt, Franklin D., 227
Rose Creek (tributary of Little Blue River), 34, 42
Rose Creek flower (fossil), 34
Rosedale Bottoms, xxv (map), 169, 170
Rosedale district (Kansas City, KS), xxv (map), 171, 186, 238
rough-legged hawk, 108–109, 110–111

ruby-crowned kinglet, 103
Ruede, Howard, 219–220
Runza® sandwich, 60
"Rural Kansas is Dying" (Brown), 176
Russell, KS, xxiv (map), 224
Russell County (KS) Courthouse, 224
Russian olive, 45
Rydjord, John, 98

Sacred Red Rock, 33. See also *Iⁿ'zhúje 'waxóbe*; Shunganunga boulder
Sage, Rufus, 84
Saint Andrew Episcopal Church, 225
Saint Anthony Catholic Church, 224
Saint Catherine Catholic Church, 225
Saint Francis of Assisi Catholic Church, 224
Saint Joseph Catholic Church (Damar, KS), 225
Saint Joseph Catholic Church (Liebenthal, KS), 224
Saint Joseph Catholic Church (New Almelo, KS), 226
Saint Jude Thaddeus, 245, 246
Saint Mary's Catholic Church (Ellis, KS), 225
Saint Mary's Catholic Church (Kansas City, KS), 218
Saint Vincent DePaul Catholic Church, 221
Salina, KS, xxiv (map), 14, 17, 183, 222
Saline River, xxiii (map), 3, 17, 21, 41, 80, 82, 85, 95, 128, 163, 225
"Saline River Country in 1859, The" (Mead), 163
sand bluestem, 61, 73
Sand Creek (CO), attack of Cheyenne and Arapahoe encampment on, 48–49
Sand Creek Massacre National Historic Site, xxiv (map)
sand dunes, 11–13
   blowouts in, 12, 61
   soil-water relationship in, 73
   See also Imperial Dune Field; Lincoln County Dune Field, Nebraska Sandhills; Wray Dune Field
sandhills muhly, 61
sand lily, 65, 66, 67
sand milkweed, 145
Sandoz, Mari, 77, 78, 161, 164

sand sagebrush, 60–61, 106
sandsage prairie, xvi, 12, 59–61, 62, 72–73, 106, 133, 134, 239, 240
sand shiner, 133
Sandzén, Birger, 23–24
Santa Fe Trail, 50, 165, 222
Sappa Creek (tributary of Republican River), 3, 52
savanna, characteristics of, in Kansas City area, 38–39
Saving the Prairies (Tobey), 61–62
Savory, Allan, 239–240
scarlet globe-mallow, 94, 212
scarplands, 25, 26. See also Flint Hills; Osage Cuestas
scarp woodland, 48
Schaeffer, Francis, 176
Scheffer, Theodore, 88, 95, 96, 97
Schmidt, Menno, 111–112
Schoenchen, KS, 224, 230
Schultz, Bertrand and Marian, 32–33
Scott County Pueblo site, 6
Scott riffle beetle, 6
Sehnert's Bakery & Bieroc Café, 182
semipalmated plover, 116, 117
semipalmated sandpiper, 117
Serda, Daniel, 171, 186
Shawnee (tribe), 79
Shellrock Limestone, 219, 223
Sherman, Althea, 100
shorebirds, 113–120, 123–124, 156, 237. See also Rainwater Basin
shortgrass prairie, 51, 52, 58, 65–67, 172, 244
   bison and, 58–59, 88–89
Short History of Progress, A (Wright), 35
Shortridge, James, 55, 168–169, 187
shovelnose sturgeon, 126
Showalter family, cemetery lilac of, 193–195
Shunganunga boulder, 33, 35. See also *Iⁿ'zhúje 'waxóbe*; Sacred Red Rock
Shunganunga Creek, 33, 131
Siberian elm, 45, 108, 178
Signal Oak, 28
silky milkvetch, 151, 152, 155
silky prairie-clover, 72–73
Silverdale limestone, 206, 222. See also Fort Riley Limestone

silver maple, 38, 40
Sioux (tribe), 84. *See also* Lakota
Sioux quartzite, 31, 33. *See also* glacial erratics
skylarks, 104–107. *See also* Cassin's sparrow, Eurasian skylark; horned larks; lark buntings
Slaughter of the Buffalo on the Kansas Pacific Railroad (Griset), 164
slender madtom, 131
Smoky Hill buttes, 14, 17, G2 (photo)
Smoky Hill Chalk, 21–23, 24, 97, 98, 200, G3 (photo)
　badlands, 21, 34, 109, 153 (*see also* Goblin Hollow; Hell's Half-Acre; Little Jerusalem)
　fossils discovered in, 34–35
　use of, as structural stone, 225, 226
Smoky Hill River, xxiii (map), 7, 83, 84, 87, 177, 183
　confluence of, with Republican River, xviii, xxiii (map), 40, 232
　headwater streams of, 3–4 (*see also* Hackberry Creek; Ladder Creek; Saline River; Solomon River)
Smoky Hills (physiographic region), xxiii (map), 14–24, 57, 62, 176, 186, 212, G2 (photo), G3 (photos)
Smoky Hill silicified chalk, 200, 202
Smoky Hill Trail, 21, 77
Smoky Valley Ranch, 98, 225
Smyth, B. B., 145, 146
snow goose, 114–115
Social History of the State of Missouri, A (Benton), 169
soil
　erosion, 172–173, 174
　Harney Silt Loam (State Soil of KS), 11
　Holdrege Silt Loam (State Soil of NE), 11
Solomon River, xxiii (map), 3, 21, 41, 42, 80, 81, 82, 85, 95, 211, 225
　South Fork of, 24, 226
southern redbelly dace, 130, 131, G7 (photo)
spadefoot, plains, 157–158
Spanish Diggings, 199, 200, 201

Spotted Tail, Chief, of Brulé band of Lakota (tribe), 83
Springsteen, Bruce, Super Bowl commercial featuring, set in US Center Chapel, 16
Stanford, Dennis, 197
star-grass, 50, 51, 62
Steele City, NE, xxiv (map), 17, 23
Steele City Canyon, 17
Stegner, Wallace, 231–232
Sternberg, Charles, 34
Sternberg, George, 34, 35
Sternberg Museum of Natural History, 34
Stevens, William Chase, 69, 138, 145
stilt sandpiper, 117
Stockville, NE, xxiv (map), 244
stonecat, 131
Stone Church (Hitchcock County, NE), 226
Stranger Creek, 32
Strawberry Hill (Kansas City, KS), xxv (map), 26, 27, 36, 218, 229
suckermouth minnow, 133
suitcase farmers, 171–172. *See also* wheat
Superior, NE, 22
Swainson's hawk, xvi–xvii, 108, 109, 243–244, 246
Swengel, Ann, 141
Swenk, Myron, 167
swift fox, 90, 97, 166
Swineford, Ada, 20
switchgrass, 51, 54
sycamore, American, xix, 38, 40, 210

tallgrass prairie, 29, 39, 51, 52, 53–55, 62, 67–69, 70, 73, 85, 130, 210, G2 (photo)
　in Kansas City area, xxv (map), 39, 50–51, 55
　restoration of, 235, 236–238
Tallgrass Prairie National Preserve, 30, 54, 131, 132, 142
Tennessee warbler, 102
Texas horned lizard, 93
threadleaf sedge, 57, 110
timber phlox, 64–65, 69
"Toad-a-Loop," 170
Tobey, Ronald, 61

Tomanek, Gerald, 136
Tomelleri, Joseph, 136
tool stone, 199, 200. *See also* Paleoindian lithic resources
Topeka, KS, xviii, xxiv (map), 33, 40, 82, 130, 131, 186, 187, 218, 232
Topeka purple coneflower, 141
Topeka shiner, 131
Torrey, John, 70, 71
Town of Kansas, xxv (map), 28, 71
Townsend, John Kirk, 104
trilliums, 38
Trinity Lutheran Church, 205
Turkey Creek (tributary of Kaw, Kansas City metro), xix, xxv (map), 170
two-spotted bumblebee, 140

University of Kansas, 48, 55, 67, 150, 221
    Memorial Campanile bell tower, 218
    Mount Oread and, 28, 221
    Natural History Museum, 35, 221
University of Nebraska, 56, 67, 150, 167, 179, 183, 231
    "Grassland School" of, 61–62
    State Museum, 32, 35, 42
upland sandpiper, 113, 117, 123
U.S. Army Corps of Engineers, 127, 128
U.S. Center Chapel, xxiv (map), 16
U.S. Fish and Wildlife Service, 97, 98, 124
U.S. Geographical Center Monument, 16
U.S. Hub Club, 16

Velez, Alex, 13
Venango, NE, xxiv (map), 13
Vermillion Creek, xviii
Victoria, KS, 205, 206, 224
Views of Nature (Humboldt), 1
Volga River region of Russia, immigrants to Kansas from, 205, 224
Volland Foundation, 131

Wachiska Audubon Society, 54
Wakarusa River, xviii, 128
WaKeeney, KS, 83, 95
Waldo district (Kansas City, MO), xxv (map), 55

Walker, KS, 224, 228
walnut, black, 38, 40, 210
Wamego, KS, 32, 33
warblers, 101–103, 131
Ware, Eugene F., 45–46
Warren, G. K., xviii
Waterkeeper Alliance, 232
Wauneta, NE, 7, 10
Weaver, John Ernest, 52, 60, 62, 67, 231
Wedel, Waldo, 79, 85–86, 87, 196, 197, 204
Wells, Philip, 48
West, Elliott, 46, 84
West Bottoms, xxv (map), 26, 169, 187
Western Hemisphere Shorebird Reserve Network, 116, 118, 123
Western Interior Seaway, 15, 16, 19, 21, 34
western kingbird, 100, 101, 107, 108, 177, 178
western meadowlark, 93, 105, 106, 108
western sandpiper, 117
Westfall, Tom, 202
Westmoreland, KS, 220
West of Wichita (Miner), 209, 215
Westport, MO, xxv (map), 28, 67, 68
Westport Landing, xxv (map), 28
wheat, 13, 29, 171–172
Where the Buffalo Roam (Matthews), 182
white-lined sphinx, 146
white oak, 28, 38, 40
White Rock Creek, 203
white-rumped sandpiper, 117
Whitney, Bill and Jan, 235–238, 241. *See also* Prairie Plains Resource Institute
Wichita (tribe), 179
Wilber, NE, xxiv (map), 177
wildflowers, 63–74
    associated with Ogallala Formation outcrops, 151–152, 153
    mixed-grass prairie, 57, 69–72
    sandsage prairie, 72–73
    shortgrass prairie, 58, 65–67
    tallgrass prairie, 67–69, 70
    woodland, 64–65
Wildhorse Creek, 24
Willa Cather Memorial Prairie, 57
willet, 119
Wilson, Edward O., 196

Wilson Reservoir, 17, 128
Wilson's phalarope, 117, 119
Wilson's snipe, 116
Wishart, David, 96, 179–180
Wishart, Lourene (and Joe), 192–193
Wislizenus, Friedrich Adolph, 68, 69
Wolfenbarger, LaReesa, 118, 124
wolves, gray, 89, 165–166
wooded plant communities, 37–49
woodland. *See* bur oak woodland; cottonwood riparian woodland; scarp woodland
Works Progress Administration (WPA)
   community parks enhanced with support of, 228
   Federal Writers' Project of, 7
   Great Depression and, 17, 33, 227
   native stone structures constructed with support of, 227–230

worm-loving warbler, 102
Worster, Donald, 172, 173
Worst Hard Time, The (Egan), 173
Wray, CO, xxiv (map), 6–7, 11, 197, 207
Wray Dune Field, 11, 12, 106, 133, 239
Wright, Ronald, 35
Wyeth, Nathaniel, expedition of, 104

yellow-headed blackbird, 104
yellow-rumped warbler, 102
yellow warbler, 102
York, NE, xxiv (map), 69, 191
Yost, Jeff, 184–185. *See also* Nebraska Community Foundation
Young Girl with Book (Felton), 204, 206–207
yucca, plains, 147–148, G4 (photo)
yucca moth, 147–148

Zimmerman, John, 113, 117